Surveillance, maintenance and diagnosis of flood protection dikes

A practical handbook for owners and operators

Patrice MÉRIAUX and Paul ROYET

Photographic and illustration credits

Photographs with captions that mention the name of a person or organisation were not the work of this handbook's authors:
– Cemagref (Cyril Folton, Rémy Tourment)
– DDE 66 (Éric Josse *et al.*)
– DDE 13
– ONF Alpes Maritimes
– Service Navigation Rhône-Saône
– Photos 1.3 (page 30) and 2.1 (page 34) appear courtesy of and thanks to professional photographer Frédéric Hédelin.
All other photos were the work of the authors (Patrice Mériaux and Paul Royet).
Illustrations designed by Cyril Folton.

Translation credits:
Aquitaine Services Logistique, Lormont, France (www.asl.fr) – Julia Rubin
Special thanks to Cathy Ricci, graduate in English, for translation review.

Surveillance, maintenance and diagnosis of flood protection dikes:
A practical handbook for owners and operators – Patrice Mériaux and Paul Royet
© Éditions Quæ 2007. ISBN 2-7592-0037-5. Registration of copyright – 3rd quarter 2007. Editing – Julienne Baudel. Computer graphics and photographic processing – Françoise Peyriguer and Joëlle Veltz. Distribution Quae.com – Price: 42 €

TABLE OF CONTENTS

Table of contents

FOREWORD

••• Objectives and content of this handbook

This technical handbook is intended for the personnel of companies or organisations involved in the management of dikes designed to provide protection against flooding caused by a rise in river levels. Written for engineers and technicians, its aim is to increase understanding of:
- How dike systems work.
- The risks faced.
- Surveillance operations.
- Maintenance operations.

It also aims to describe and explain the work to be done to ensure the long-term future and safety of such structures, in view of water resource and flood prevention legislation in France.

Produced upon the initiative of the French Ministry of the Environment (water resources department), this book was written by Cemagref, under the guidance of, and with contributions from, a working group.

Working group members:

– Sébastien de Bouard	Highways department (CGPC)
– Gérard Couzy	Highways department (CGPC)
– Jean Varret	Department of agricultural engineering, water & forestry resources (CGAAER)
– Jean-Jacques Vidal	Regional Ministry of the Environment (DIREN) – "Midi-Pyrénées"
– Jean-Noël Gauthier	"Plan Loire grandeur nature" multidisciplinary team
– Zbyniev Gazowski & Didier Reinbold	Regional department of the Ministry of the Environment (DIREN) – "Centre"
– Michel Lescure & Jean-Michel Colin	"Gard" town & country planning office (DDE)
– Pierre Le Floch	"Indre-et-Loire" town & country planning office (DDE)
– Yannick Fagon	Centre for Maritime & Inland Waterway Studies (CETMEF)
– Jean Kloos	"Lot-et-Garonne" town & country planning office (DDE)
– René Feunteun	French Ministry of the Interior (DDSC)
– Philippe Pipraud	French Ministry for Agriculture & Fisheries (DERF)
– Jean-Michel Tanguy	Mediterranean technical studies & engineering centre (CETE)
– Jean-Luc Roy & Marie-Pierre Nerard	Ministry of the Environment – water resources department

This book forms part of a nationwide French initiative to improve the safety of flood protection installations the failure of which would have serious repercussions for both people and property (dikes that pose a potential risk to public safety).

To this end, the French government has introduced a scheme (see Appendix 5) to control:
– Practical measures implemented by operators.
– The safety of such structures.

It is for this reason that the last chapter of this book describes one method of dike diagnosis.

Layout

Following a description of dikes and their functions in Chapter One, Chapter Two considers the various malfunctions and failure mechanisms that may affect these structures.

Visual inspection, upon which diagnosis and surveillance are based, is dealt with in Chapter Three. Appendix 3 contains a methodology for recording information obtained from visual inspections, together with standard anomalies record forms.

Specific aspects of dike surveillance during flooding form the subject of Chapter Four.

Chapter Five deals with the maintenance of dikes and appurtenant works; it provides practical advice on the most common repairs.

Finally, Chapter Six briefly describes the stages in dike diagnosis.

A separate French publication provides further information on this subject: *"Métho-dologie de diagnostic des digues appliquée aux levées de Loire moyenne"* (methodo-logy of dike diagnosis applied to levees along the middle reaches of the River Loire, March 2000, Cemagref Éditions).

This handbook also contains a short list of terms specific to dikes (see Fig. 1 also), including explanations of abbreviations used in the book, as well as the basic prin-ciples of soil mechanics (Appendix 1) and soil hydraulics (Appendix 2). Finally, refe-rences to prices are given in euro (excluding taxes).

Appendix 5, which briefly outlines French legislation on flood protection dikes in the last decade, has been added to the 2004 edition.

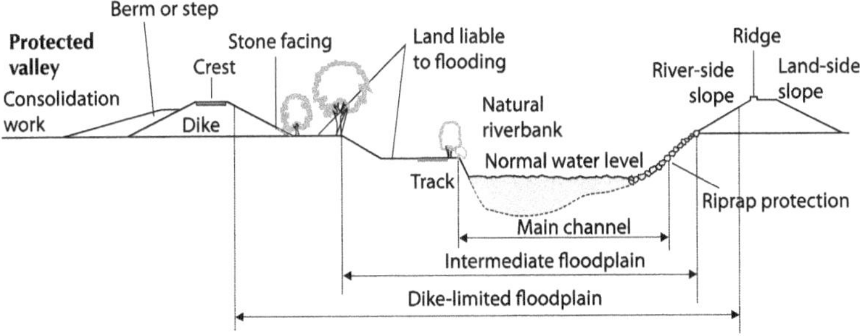

Figure 1. Typical cross-section of diked land

Roles and responsibilities of those involved in dike management

Several categories of interested parties or players are more or less concerned by dike management. The initial difficulty lies in identifying them. To this end, we propose the following list (see Fig. 2):

- ***Dike owner:*** This is usually the dike builder and may be the State, a local commu-nity or group of local communities, a property owners' syndicate, private individuals, etc.

- ***Site owner:*** The owner of the foundations upon which the dike stands. More often than not, the site owner is also the dike owner (the ideal situation), but they may be different entities whose relationship is not always clearly defined.

- ***Owner or operator of structures or networks*** built on or into the body of the dike, including buildings, functional premises, electricity or telephone lines, gates and stop log structures, culverts, conduits and pipes, communications channels, etc. It is a very good idea for a written agreement to be signed with the dike owner and/or operator, in which each party's responsibilities are specified.

– *Owner or manager of land and/or constructions (liable to flooding)* situated between the dike and the main channel of the river or stream.

– *Owner or manager of land and/or constructions* (valley side, land side) protected by the dike, but exposed to the risks of failure or of flooding in the wake of failure.

– *Dike operator:* When the owner and operator are different entities, the owner makes the operator responsible for the maintenance and correct working of the dike, in principle by way of a formal agreement.

– *Director of works:* In charge of dike building, heightening and upgrading. Logically this is the dike owner, but it is possible for an institution to assume the role of director of works for structures not belonging to it (e.g. a group of local communities in charge of works on private land or structures).

– *Design office* or *engineering office* (private or public), under contract from the owner, director of works or operator to carry out preliminary research (diagnosis, design, consultancy, etc.) or to supervise work done on the dike.

– *Company:* Responsible for the construction, heightening or upgrading of a dike.

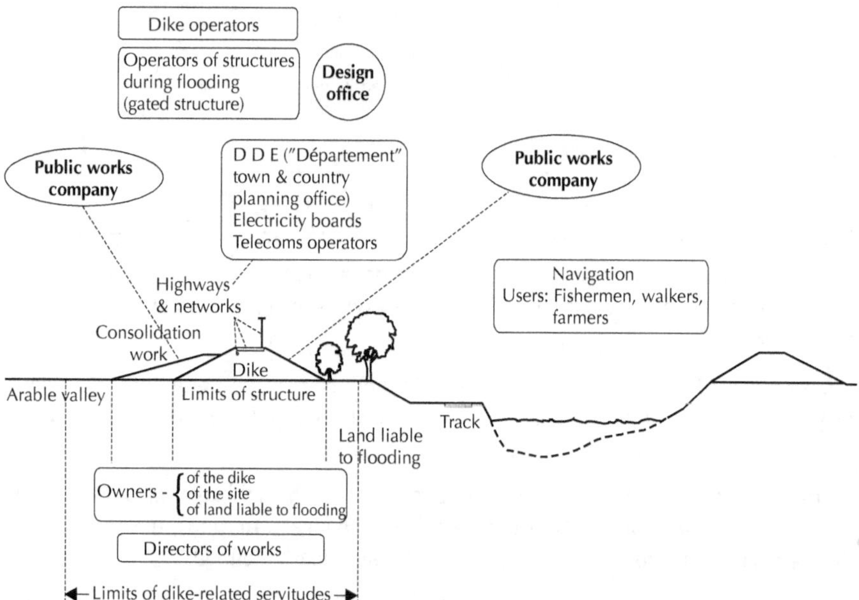

Figure 2. Responsibility for the safety of people and property

- **_French Water Police:_** In charge of controlling authorisation procedures relating to structures and of verifications of dikes identified as structures that pose a threat to public safety.

- **_Official bodies responsible for the safety of people and property:_** local and regional authorities, local councils.

RESPONSIBILITY OF THE OWNER

The dike *owner* (local community, group of local communities, a property owners' syndicate, private individual, etc.) is wholly responsible, under both civil and criminal law, for any damage that may be caused by the structure, and, in particular, by its failure.

This responsibility may, in principle, be reduced in certain situations (major floods classed as natural disasters or considered to be unforeseeable). On the other hand, an obvious failure to correctly monitor and maintain structures is likely to worsen circumstances.

THE NEED TO MONITOR AND MAINTAIN DIKES

Responsibility apart, the objective of keeping structures in good condition provides sufficient justification for regular surveillance and maintenance for two main reasons:
- Regular surveillance means that a great many anomalies and malfunctions can be detected at an early stage, that subsequent developments can be tracked and that any necessary maintenance and repair work can be carried out in good time.
- If a structure is properly maintained, it will age more slowly and have a longer service life. For instance, routine work to clear invasive ground cover or dissuade burrowing animals can do away with the need for more substantial rehabilitation work. Good maintenance of structures (especially control of vegetation and maintenance of service tracks) also makes surveillance and inspection easier.

TECHNICAL SKILLS REQUIRED FOR STRUCTURE MANAGEMENT

To fully assume the role described above, the *owner* of a dike system needs field technicians who have been trained to carry out the various tasks associated with surveillance and maintenance. If these technicians work directly for the owner, the latter is also considered to be the *operator*. The supervisors of operating and maintenance workers should also be conversant with geotechnics, civil engineering, hydraulics and environmental engineering. Therefore, and if need be, it is recommended that owners sign formal agreements for the management of dikes with organisations that have suitably qualified staff or technical departments that are capable of operating a stand-by system in the event of an emergency.

In this case, the owner is distinct from the operator, but their contractual ties should be clearly defined.

It is therefore recommended for small entities that own dikes to entrust their management (or even transfer ownership) to appropriately-sized organisations that have the resources needed for good management and operation.

CLASSIFICATION OF DIKES: RECENT CHANGES TO FRENCH REGULATIONS

Since 2002, flood-protection dikes in France have been governed by a system of authorisation or declaration, depending on their size (*cf.* decree dated 13/02/2002 mentioned in Appendix 5). They may be classed as being "a potential threat to public safety" if their failure would result in a serious risk to human life (*cf.* circular dated 06/08/2003 referred to in Appendix 5).

1

Nature, functions
and composition of dikes

DIKE: An artificial structure, flood protection embankment or barrier that protects against river flooding, at least a part of which is built above natural ground level. It is designed to periodically contain a high discharge of water and thus protect areas that are naturally prone to flooding. (The term "levee" is often used along the French River Loire in common with certain areas of the USA).

Simple constructions that protect the slopes of riverbanks (masonry walls, riprap or concrete facings) but that are no higher than the top of the natural bank are not considered to be dikes. Neither are quay walls, unless incorporated into a dike in the above-mentioned sense of the word, nor structures intended to protect against coastal erosion (groynes, seawalls, etc.) or harbour jetties.

This handbook does not cover:
– Canal embankments (navigable waterways, hydroelectric plant feeder canals, etc.).
– Highway and railway embankments situated in floodplains.
– Bank protections not topped by a dike.

We should also mention "sea dikes", the function of which is to protect estuaries and coastal areas against high tides or unusually high seawater levels created by storms, as in the Camargue, for example, at Salins-de-Giraud. A French guide to such structures is to be published at a later date.

1.1 Overview of existing structures in France

Though not very well known, France has a considerable number of flood protection dikes. It is generally only during major floods that they make the news headlines, when failure leads to the flooding of supposedly protected areas.

A national enquiry, initiated in 1999 by the French Ministry of the Environment with a view to compiling a complete evaluation and survey of these facilities (creation of a DIKES database of structures, operators and potential consequences of failures), led to the initial observation that the country has some 8,000 kilometres of dikes and a thousand or so operators. To mention just some:

– Along the 450 km of the middle reaches of the River Loire (between the confluences of the Allier and Maine rivers), 600 km of mainly state-owned dikes (known locally as "levees") protect 1,000 sq.km or so of valleys liable to flooding. To this should be added the levees built along tributaries such as the Rivers Cher, Indre and Vienne. Several large towns are protected by levees, including Tours, where 90,000 people are concerned, Orléans and its urban area with 40,000 inhabitants, Blois with 10,000 and the Authion flooding valley, close to Saumur, with 45,000. The Loire levee system has not been subject to major flood peaks since the three great floods of the middle of the 19th century, the consequences of which were considerable.

– The course of the River Garonne (in southwest France) was extensively diked following flooding in 1875, when 500 people lost their lives, including 200 in Toulouse. The dikes did not, however, prevent the loss of another 200 lives in the 1930 flood. Although over 90% of the land liable to flooding and protected by dikes alongside the Garonne is agricultural, the populations of a number of large towns are still directly at risk, including 40,000 people in Toulouse and 25,000 people in Agen. The status of dikes along the Garonne varies widely.

– Along the two branches of the Rhône delta, the Camargue is protected against flooding by approximately 200 km of dikes, which were breached in 16 places during flooding in October 1993 and January 1994 (floods considered to be hundred-year events). These breaches were largely due to deficient dike surveillance and maintenance; the management system (small property owners' syndicate) was acknowledged to be inappropriate and has since been substantially modified.

1.2 How dike systems work (dikes and spillways)

The hydraulic behaviour of a dike-limited floodplain can be described as follows:
– During a flood, rising water levels cause the river to break out of its main channel and spread over into the diked floodplain (Fig. 3a).

– A dike system limits the spread of flood waters during minor and medium-intensity floods, but it also leads to a rise in water levels at the point where it reduces the width of the natural river bed (a common feature in urban areas).

– Flood peak reduction (which attenuates maximum discharge by propagation into parts of the floodplain) is thus limited during regular floods.

– Areas protected by dikes may, in certain cases, be flooded by main river water backing up into a tributary, by runoff from lateral catchment basins whose outlets into the river are saturated or by a rise in the water table (Fig. 3b).

– To prevent overtopping (and the virtually certain failure) of dikes during a major flood, spillways are sometimes built into them which, when the water exceeds a certain level, make it possible to flood areas that are less built-up, giving flood waters more room to propagate and thus facilitate discharge (Fig. 3c).These deliberate flood propagation areas are sometimes themselves divided by embankments into a number of flood spreading plains that are inundated in turn.

– In an extreme flood, the whole valley is inundated, either following spillway operation or because of breaching as the result of dike overtopping. The watercourse then covers its entire floodplain as though in the absence of flood defences.

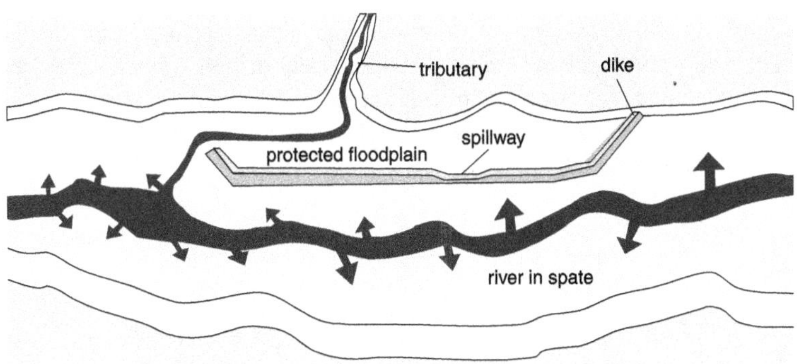

Figure 3a. Propagation of flood waters in a diked floodplain

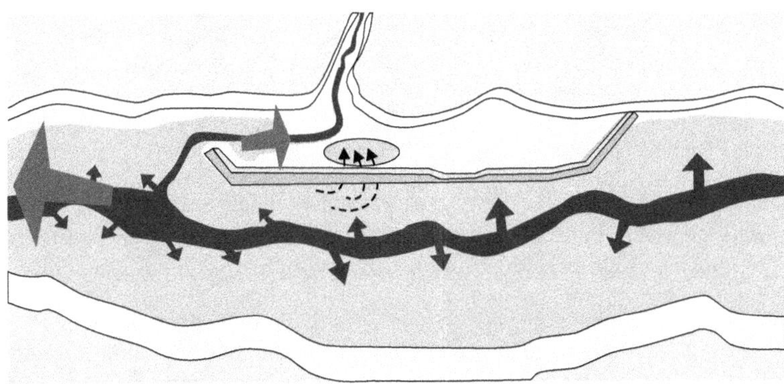

Figure 3b. Flooding of a valley by backing up, runoff from a catchment basin or a rise in the water table

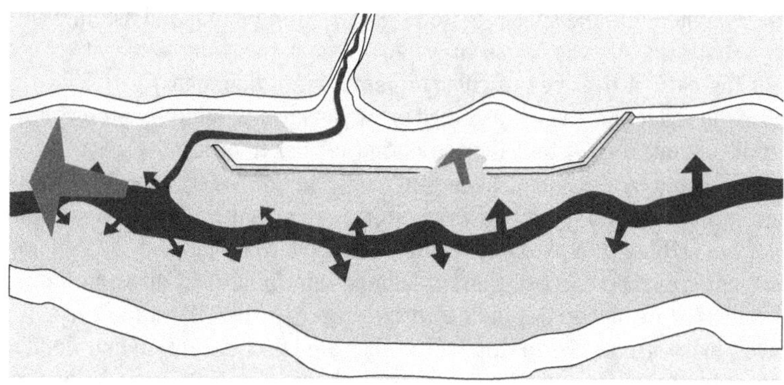

Figure 3c. Spillway operation

1.3 Composition of dikes

1.3.1 Fill dikes

The majority of dikes in France are earthfill embankment constructions made with materials ranging from silt to sand and occasionally gravel. Their composition can largely be explained by the history of their construction.
– Very often they were built in stages during different periods as the use of rivers and the need for defences changed (Fig. 4).

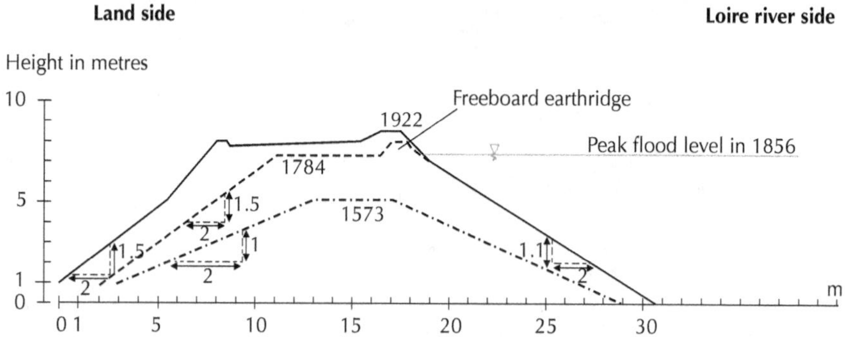

Figure 4. Typical cross-sections of River Loire levees before recent upgrading work

– Since powerful earth-moving equipment was not available at the time, dike embankments were generally built with materials taken from the immediately surrounding area; the remains of old borrow pits can still be seen at the toe of some dikes.

The nature of earthfill materials may therefore vary widely, even along the same river (sandy in the middle reaches and silty nearer the mouth). Generally, however, single sections are of a homogeneous nature with no zoning and no special internal drainage system (Fig. 5).

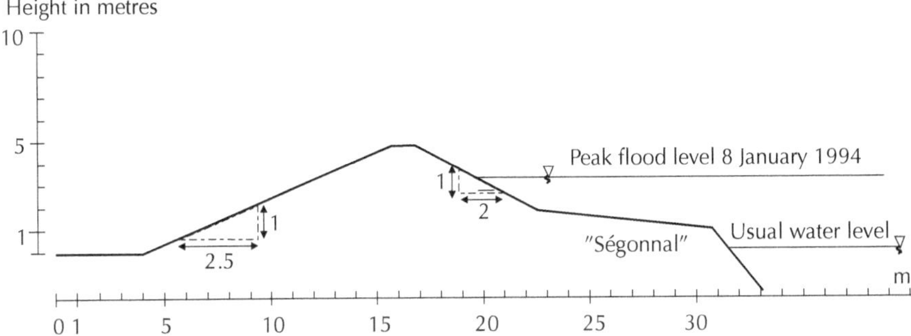

Figure 5. Typical cross-section of a dike on the Petit Rhône in the Camargue

– Likewise, the absence of powerful compacting and excavating equipment when the old dikes were built resulted in relatively poorly-compacted embankments, which were not especially well-anchored into foundations, themselves not made particularly impervious.

– Slopes are generally protected by grassing. On the river side, stone facing often protects areas in contact with the main channel, although it may be hidden by silt deposits and vegetation.

– In areas most exposed to scouring, toe protective works have sometimes been incorporated, usually made of secant wood piles.

– Efforts to increase freeboard (or safety vis-à-vis overtopping) have sometimes led to dike crests being raised by means of narrow earth ridges or freeboard masonry walls (called "banquettes" in the Loire), generally built on the river-side edge of the crest.

The most recent dikes use designs similar to those for dams, which integrate materials zoning and the separation of sealing and draining functions (Fig. 6).

1.3.2 Masonry and concrete quay walls

When the available surface area for a structure was limited (generally in urban areas), wide gravity-retaining walls were built using cut stone. This was the case in the Loire for the majority of sections passing through towns, along the River Garonne in Toulouse and Bordeaux, on the River Rhône in Arles and in many other sectors.

These walls are characteristically steep on the river side and are often supported on the land side by an embankment, made of earth or coarse materials, which may also be topped by a road (Fig. 7). More recently, concrete has replaced stonework, although an external facing of stone is occasionally re-built to make a more attractive finish.

In some areas, gabion structures have been used to protect the river-side facings of fill dikes.

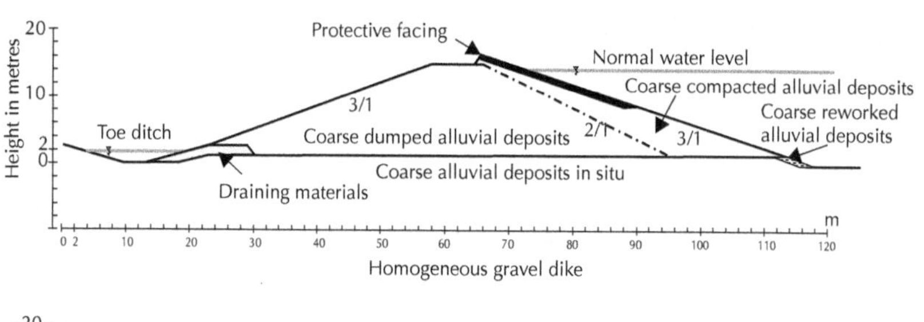

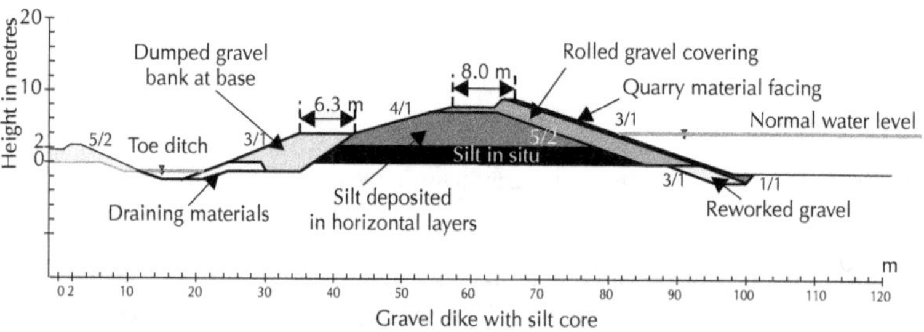

Figures 6. Typical cross-sections of Rhône dikes in the sector developed by the Compagnie Nationale du Rhône (CNR)

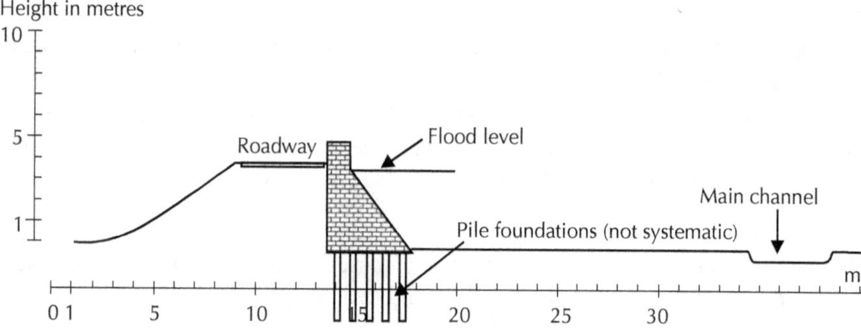

Figure 7. Typical cross-section of a masonry dike

1.3.3 Spillways

Dikes are not designed to contain exceptional floods (typically occurring every hundred years or more). In some cases, to anticipate the dangers of overtopping, which would almost certainly lead to sudden failure, spillways have been added, the top of which is built a few dozen centimetres (traditionally about a metre in the case of Loire spillways) below the dike crest. They are designed to allow flood waters to spread into what is, in principle, a low-risk containment area, thus preventing overtopping (and any ensuing dike failure).

Spillways may simply be low points dug out of the natural terrain, but more frequently take the form of a sill with a dressed stone shell that covers the fill material.

A discharge chute leads from the sill down the land-side slope to dissipate the energy of the flowing water.

A masonry sill may sometimes be topped by an erodible earth ridge (earthfill fuse), lying slightly below the dike crest. This erodible fuse is designed to quickly wash out as overtopping begins, leaving a larger section free for flood routing (Fig. 8). This effectively delays the moment at which the spillway starts to function and valley flooding is less frequent.

The most recent spillways are made entirely of concrete.

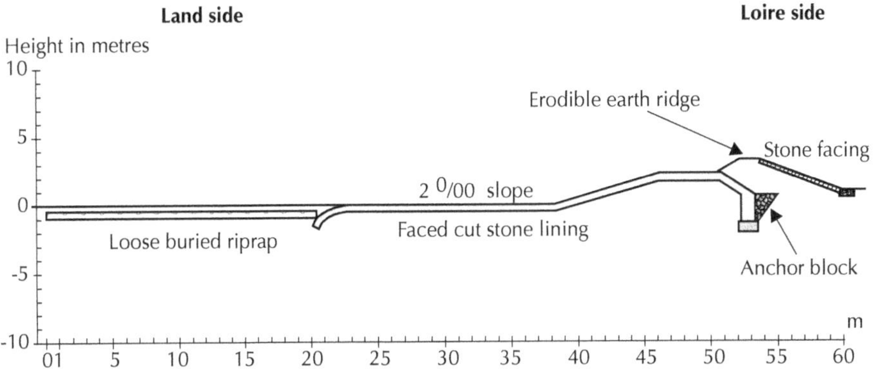

Figure 8. Typical cross-section of a "Comoy" spillway on the Loire river – (Ouzouer)

1.3.4 Particular structures and features

Dikes are linear structures, but their longitudinal homogeneity is frequently interrupted. Many particular structures and features have been built into or across them, either as part of dike development or as compensatory measures. These include stop log structures and ramps providing access to the river, through-dike aqueducts, tunnels, culverts, conduits and pipes, whether equipped with flap valves or not.

In some cases, buildings have actually been embedded into dikes, either originally or during subsequent upgrading work. Take for example the houses that Henry Plantagenet II had built into the levees of the Authion valley (River Loire) in the hope that those who occupied them would make sure the dikes were properly maintained since they were the first to be affected by their state of repair(!).

2 CLASSIFICATION OF MALFUNCTIONS AND FAILURE MECHANISMS

Of the various mechanisms that can lead to dike failure, special mention should be made of overtopping since it is caused by a typically external phenomenon, i.e. a water level that rises higher than that of the *datum event* that was used to determine the height the dike should be built. It is, therefore, essential to have information about that datum event in order to work out the level of protection afforded by the structure. This stems from hydrological and hydraulic studies. In contrast with this, all other failure mechanisms described below relate directly to the strength of the dike and are, therefore, closely associated with its location, geotechnical design, surveillance and appropriate maintenance.

Finally, it is important to note that the various degradation mechanisms described below (§ 2.1 to 2.4.) may act simultaneously or sequentially in a process culminating in the failure of the dike. For example, erosion of the river-side slope increases the likelihood of internal erosion in the backfill if the transverse section of the dike has been significantly worn away, or may lead to sliding on the river-side slope (since the overall gradient of the slope is steeper) when water levels drop after a flood peak.

2.1 Overtopping

Overtopping (when water rises above and flows over the crest of a dike) generally and rapidly[1] causes breaching (in the case of fill dikes), by retrogressive erosion of the land-side slope and then of the crest (Fig. 9). It is one of the principal (if not the principal) mechanisms identified in the failure of fill dikes, at least if we consider the major incidents that have affected French inland dikes during serious floods in the last two hundred years.

Start of overtopping:
The water level reaches the crest of the structure.
Water flows over the dike and down onto the floodplain.

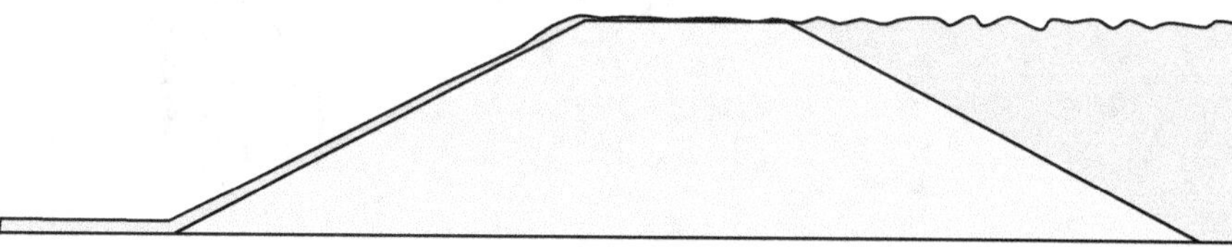

A few minutes later:
The downstream facing begins to erode. Materials
are torn away by the force of the current at the toe of the dike.

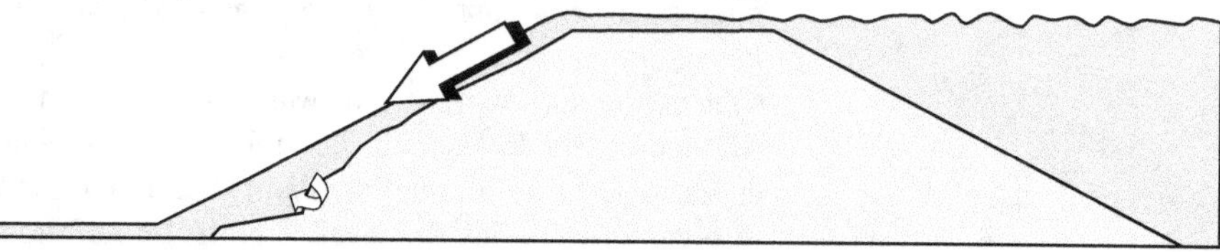

The dike facing is badly broken up,
the toe of the slope is seriously undermined
and the structure is soaked with water.

Figure 9. Mechanism of dike failure caused by overtopping

---

1. Although it has been known for dikes that are protected by regular grassing to have withstood being overtopped by a few centimetres of water for several tens of minutes.

The waterlogged downstream facing is no longer stable and whole sheets of it slide downwards. Materials are quickly washed away by the current, which is getting faster.

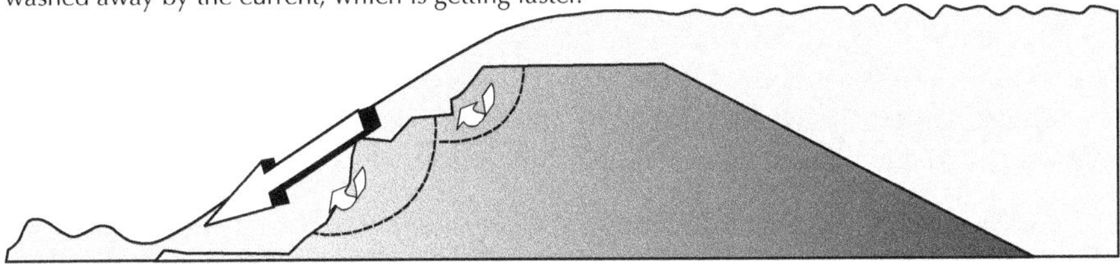

The process of disintegration accelerates, materials are torn away by the force of the current, leading to total destruction.

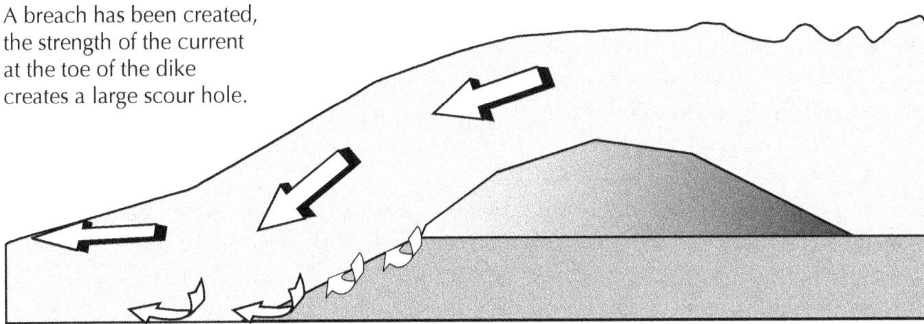

A breach has been created, the strength of the current at the toe of the dike creates a large scour hole.

The breach widens as it erodes the dike on both sides.

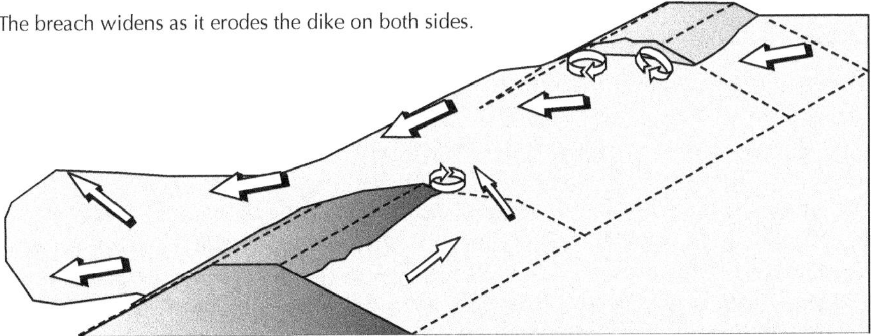

Figure 9. (*cont'd*)

Overtopping of the Loire levees was responsible for nearly half the breaches that occurred during the three floods of the mid-19th century, without counting overtopping due to water flowing back to the river. In over two-thirds of the cases of overtopping, a low point in the dike's longitudinal profile was identified as the cause of the failure, since it concentrated the flow of water over the dike at this point. Rises in water levels on the concave banks of bends in rivers or upstream of bridges or piers obstructed by logs or debris were also points at which overtopping occurred.

In general, not enough accurate data is available to gauge the height and duration of overtopping that has initiated retrogressive erosion and the opening of breaches. In answer to the question, *"How much overtopping can dikes withstand?"*, because of our current lack of knowledge and as a precautionary principle, the conservative answer must be, *"Earth dikes do not withstand overtopping".*

However, this answer can be qualified as follows:
– The proportion of sand in the materials used to build embankments, together with heterogeneous compactness, seriously affects a dike's susceptibility to overtopping.
– An irregular longitudinal crest profile, with low points caused by levelling faults, differential settlement or poor quality earthworking, will encourage local concentration of overtopping flows.
– Moreover, a well-compacted dike with a very uniform longitudinal profile, well-grassed embankments and a hard-surface crest is probably likely to withstand several centimetres (perhaps more) of overflowing water for a limited period of time.

The possible presence of freeboard earth ridges does not reduce the risk of overtopping, since they tend to be narrow and poorly compacted (as in the Loire); at most, they constitute a degree of protection against waves.

Returning-flow failures

This is a particular case of overtopping when overspill water crosses back over the dike into the river further downstream. It also occurs when water arriving from a lateral tributary basin saturates its outlets and fills the valley. The retrogressive erosion that follows this overtopping incident occurs more or less rapidly depending on how many hours or days the slope has been absorbing flood waters. This causes failures that occur just when valley flooding is at its most intense: large quantities of water that were diverted further upstream now return to the river by way of the breach, which increases flooding downstream.

2.2 External erosion and scouring

During floods, the river-side slopes of dikes, as well as the banks that sometimes run directly alongside them, are subject to the effects of currents that can cause erosion at their bases. This causes a localised increase in the gradient of the slope which, combined with a weakening of the mechanical properties of the structure (due to saturation of component materials), then leads to landslides, which in turn initiate hydraulic disturbance (eddies) and erosion. If the process continues working its way back up the slope, the result may be a breach in the dike itself. Overtopping will tend to speed up this phenomenon (see Fig. 10).

High velocity of flow and river-bank vulnerability
are at the origin of erosion at the toe of the dike.

The toe of the bank deteriorates and saturated materials slide downwards.

Repeated floods exacerbate the problem; the toe of the dike continues to break up.

The bank, now vertical, is highly unstable.
Whole sections of saturated materials slide down the slope, breaking up the dike as they go.

Figure 10. The scouring process at the toe of a dike

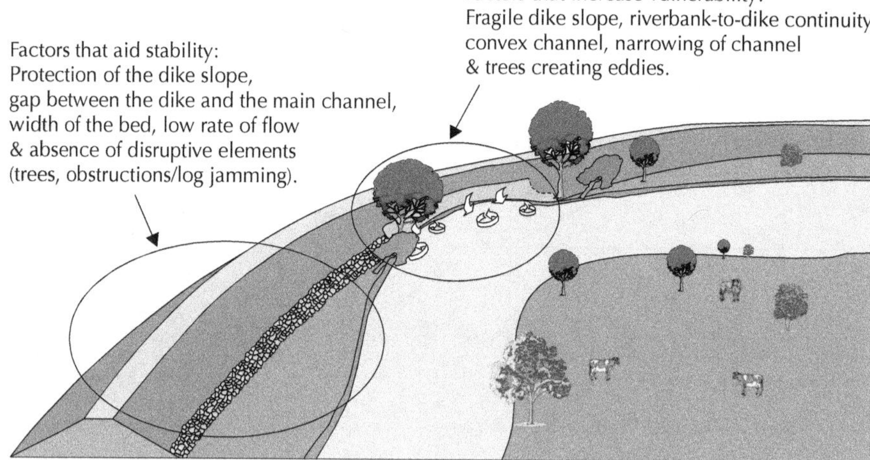

Figure 10. (*cont'd*)

Three factors increase the likelihood of this type of deterioration:
– The *average velocity of the water* travelling along the slope, which is a function of the distance of the dike from the main channel and/or river bank. Dikes built immediately next to the main channel (and directly over the river bank) are especially vulnerable, as are dikes built where the floodplain narrows.
– *Localised hydraulic disturbance* capable of generating currents and eddies with a higher localised velocity than the average for that segment. Trees, piers and any construction on the river-side facing of dikes are the source of such hydraulic discontinuity, as are pronounced bends along the dike.
– *The nature and condition of the protection* on the river-side slope of the dike. Masonry facing that is in good repair is considered to be able to withstand an average water velocity of 4 metres per second, whereas a grassed slope will only withstand a maximum of 1.5 metres per second. A change of materials along the dike slope (e.g. from stone facing to a grassed area) is also an important factor in vulnerability.

Deterioration caused by external erosion can also occur on the land side although, apart from erosion associated with overtopping (see section 2.1), this is usually restricted to areas near spillways (higher velocity at the beginning of spillover prior to valley flooding).

2.3 Internal erosion (or piping)

Heterogeneity in the permeability of a dike body (if earthfill) and its foundations (whatever its composition) can be responsible for the creation of preferred paths along which water will tend to circulate when the structure is exposed to flooding. Depending on the hydraulic head and the nature of the materials, the hydraulic gradient may locally reach a critical level, causing internal erosion and gradually creating a conduit along which gradient and velocity increase rapidly over time. If the phenomenon intensifies, the result may be a tunnel (or pipe) running across the dike or its foundations and, eventually, a breach due to caving-in (see Fig. 11).

This phenomenon was found to be the cause of sixteen breaches (4.7%) during the Loire floods of the 19th century. Over half these breaches appeared where a fill section adjoined a masonry section.

Process of piping (or retrogressive internal erosion).
Because of the rise in the upstream water level (H),
the embankment gradually becomes waterlogged.
The hydraulic gradient (H/L) increases.

A few minutes later.
Water begins to seep along preferential paths,
leading to the beginnings of a leak
on the land side of the structure.

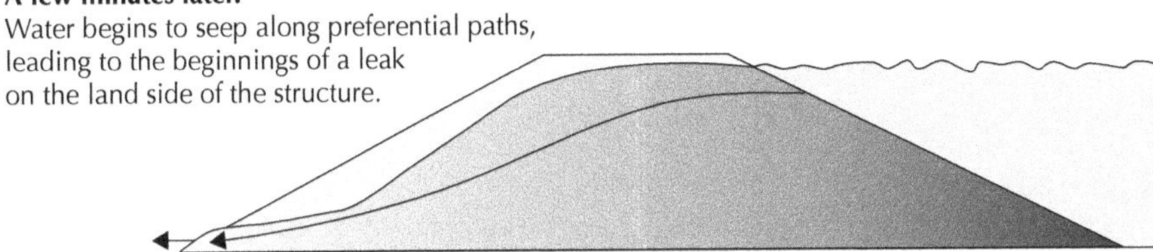

Seepage is now well under way.
Non-cohesive materials are carried along by the
water to the outlet of the leak.
The hydraulic path gradually gets shorter and the
hydraulic gradient (H/L) increases
and accentuates the phenomenon.

The leak gets bigger.
Materials carried along by the water leave a void behind them, which
grows into a cavity that travels back towards the river side and widens
on the land side of the dike.
The resulting tunnel can travel back across the entire width
of the structure and end in its destruction
in one or several successive floods.

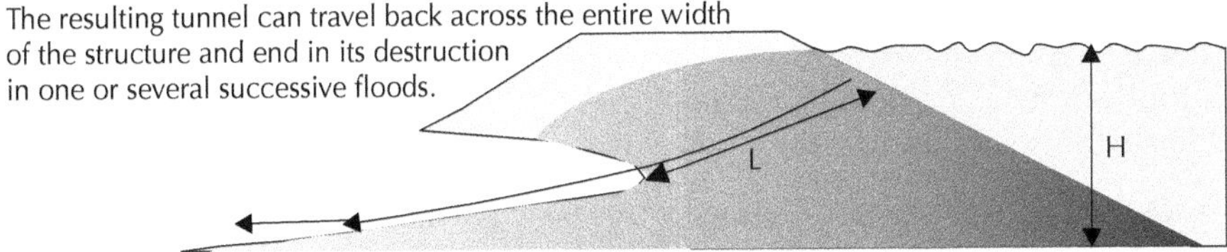

Figure 11. The process of piping

Aggravating factors.
Animal burrows and warrens and the roots of trees
form seepage paths that lead to piping.

Aggravating factors.
Across-dike installations, such as irrigation water intakes,
buried cables and evacuation tunnels can cause retrogressive erosion by
encouraging internal movement of water and seepage.

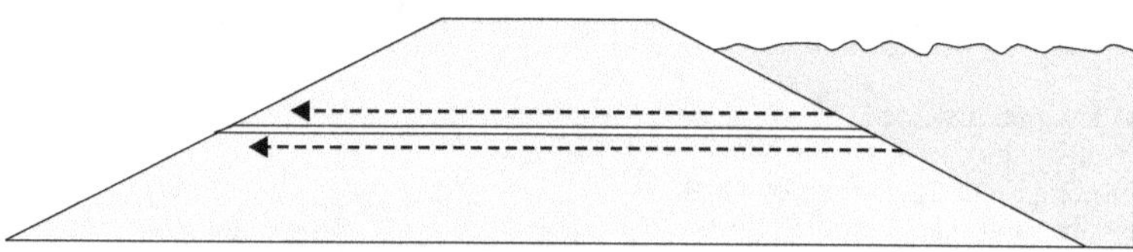

Aggravating factors.
Embankment heterogeneity, pockets of sandy materials and embedded buildings,
which facilitate internal movement of water by shortening
the hydraulic path, may cause piping, as can pervious foundations.

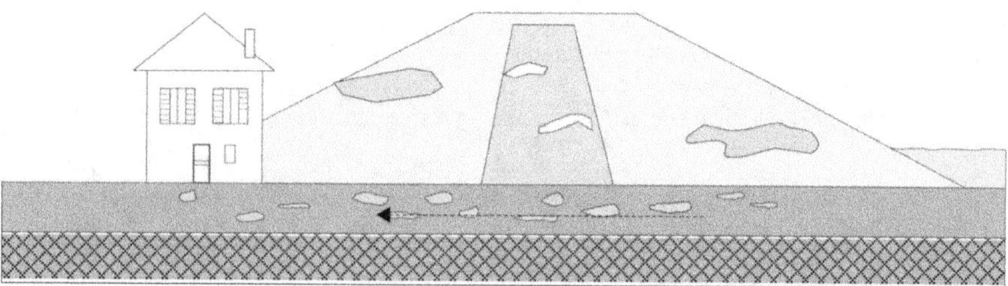

Figure 11. (*cont'd*)

During the Rhône floods of 1993 and 1994, the same type of phenomena was respon-
sible for all sixteen breaches in the Camargue embankments:
– Thirteen cases of animal burrowing.
– Three cases of pipe crossings.

Susceptibility to this kind of damage stems mainly from:
– Excavations or tunnels in the embankment, which shorten the hydraulic path
between the upstream (river side) and downstream (land side) slopes. This category

obviously includes animal burrows and warrens and the linear cavities left by decayed tree roots, as well as construction work within the dike itself.

– Poor sealing at the point where earthfill and transverse structures meet. Works built into dikes are indisputably principal risk factors here, as are culverts, pipes and tunnels crossing from one side to the other of the dike body or foundations.

– Heterogeneities in the layers of materials that make up the embankment or foundations. This risk is probably greater for foundations, which often contain alluvial deposits with a variable grain size distribution and which are seldom treated appropriately. This category also includes the sink holes that may appear in the case of karstic foundations.

2.4 Generalised slope failure

2.4.1 Fill dikes

Fill dikes typically have a transverse section whose general stability usually allows them to withstand all possible loading configurations.

Moreover, whether we consider the case of the Camargue during flooding in 1993 and 1994 or the Loire levees in the 19th century, no breach has been categorically linked to the sudden failure of a levee under load, perhaps with the exception of the Acacias levee near Blois in the Loire Valley.

However, we can consider that the risk of general instability under hydraulic load (in the form of instability on the land-side slope as illustrated in Fig. 12.1) does exist, particularly in the presence of three factors:

– Narrow dike cross-section with steeply-inclined slopes (gradient steeper than 0.65 or batters under 3 H/2 V).

– High pressure measurement (piezometry) in the dike associated with a lack of drainage and the presence of heterogeneous strata.

– Poor compactness and, therefore, fill materials with poor mechanical properties, or the presence of an under-consolidated clay-rich layer at foundation level.

These three factors are all potentially present in areas of former breaching, where repair work has not necessarily been carried out in optimum conditions.

A waterlogged embankment and a steep incline
provoke extensive failure through landsliding

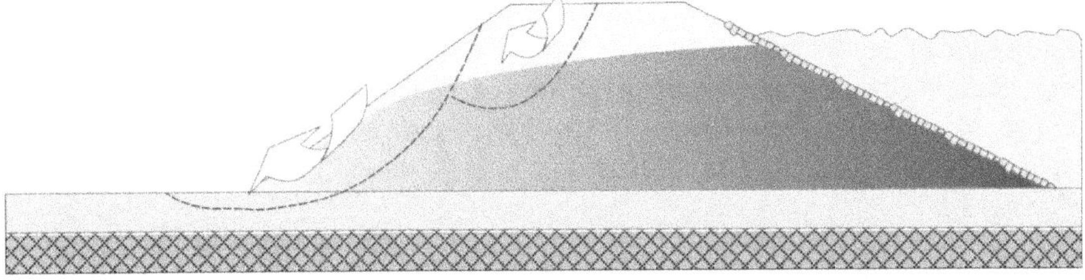

Figure 12.1. Failure mechanism on the downstream (land side) slope in a flood situation

Another type of instability occurs when water levels retreat rapidly, leading to the failure of the river-side slope and/or its protective facing (Fig. 12.2). This phenomenon, linked to uplift pressures that develop when water levels are high, particularly affects clay-rich embankments with steep slopes or protective stone facing that is too impermeable[2]. There is a real risk of this close to fuse plug spillways. Spillway activation is accompanied by rapid washing away of the fuse plug; the flow rate intercepted by the spillway increases just as rapidly, causing a drop in the river water level (see Fig. 12).

SPECIAL CASES OF DIKES TOPPED WITH FREEBOARD FEATURES

This applies to the majority of Loire levees where, following flooding in 1846, the embankments were artificially raised by the addition of freeboard ridges ("banquettes") on the river-side edge of the crest. They generally take the form of a narrow, steep-sided earth ridge (with a crest width of approximately 0.50 m and sides of roughly 0.7 m) or a masonry ridge 0.3 to 0.5 metres wide. The height of this ridge may be as much as, or even more than, 1 metre.

The join between the ridge and the existing embankment constitutes a weak point, especially from the hydraulic point of view, with the risk of preferential flows leading to erosion, piping and collapse of the ridge.

But it is also the mechanical stability of these freeboard features that appears to be poor, either because of an inadequate cross-section and the poor mechanical properties of frequently badly-compacted ridges or because of poor foundations under freeboard masonry walls, which may show signs of structural faults (cracking, leaning towards the river, etc.); phenomena that are often made worse, or even initiated, by the weight of traffic, when a dike is topped by a road.

Without distinguishing the exact mechanism involved (piping or general instability), failure of freeboard ridges or walls was responsible for 24% of Loire dike breaches in the 1866 flood. This is therefore a major source of vulnerability for dikes topped with such features, at least in segments where the ridge has not been consolidated recently.

2.4.2 Masonry dikes or parts of dikes

Such structures are especially common in urban areas (quayside walls, protection of river bank roadways, etc.) because they satisfy the constraints of space that dictate their construction. The oldest (generally not very high) are made of cut stone or non-reinforced concrete. Those that were built more recently (or are still being built) are made of reinforced concrete (e.g. free-standing pre-fabricated elements). Failure under load of this type of narrow cross-section structure (especially in the case of reinforced concrete works, the width-to-height ratio of which may be low) is necessarily sudden and may be the result of a design error (e.g. inadequate proportions) or a building fault (e.g. unsuitable reinforced concrete). The difficulty of making a diagnosis prior to failure lies in the fact that such anomalies are not necessarily apparent at the time of visual inspections.

---

2. In this respect, masonry facings are not expected to make a levee impervious and masonry joints should not, therefore, be continuous.

Failure mechanism.
When the river water level remains high during
a long-lasting flood, the embankment
becomes progressively waterlogged.

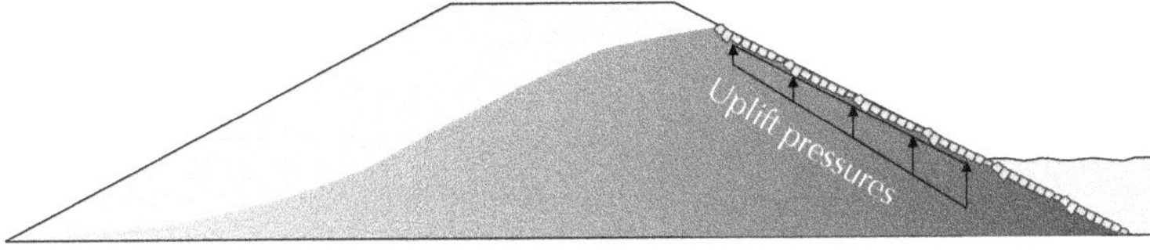

When the water level retreats.
The fall in the water level is relatively fast.
The embankment drains off more or less quickly, depending on the permeability of materials.
The facing is no longer stabilised by water pressure.
The water that has saturated the embankment changes the mechanical properties of the structure,
creating uplift pressures where the facing is badly drained.

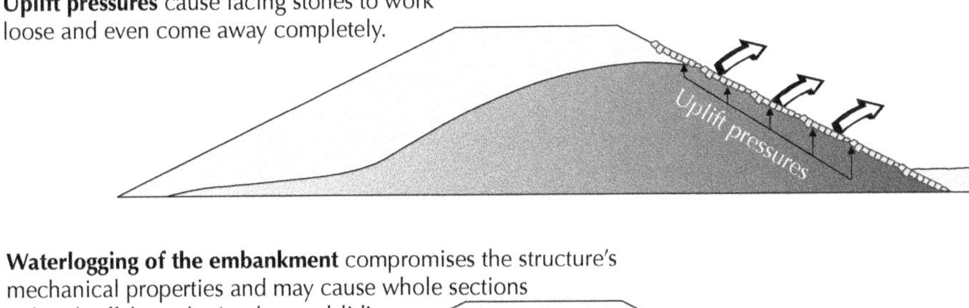

Figure 12.2a. Example of a masonry-faced slope

Uplift pressures cause facing stones to work
loose and even come away completely.

Waterlogging of the embankment compromises the structure's
mechanical properties and may cause whole sections
to break off through circular mudsliding.

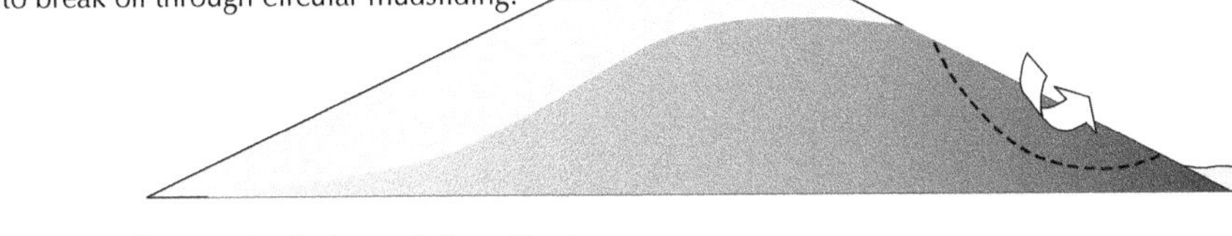

Figure 12.2b. Example of a 'natural' river-side slope

Aggravating factors:
– Very permeable and poorly compacted materials, which contribute to embankment
saturation.
– Steep gradient of facings.
– Absence of drainage, notably in protective masonry facing.
– An unstable and unconsolidated foundation layer, which increases the risk of soil
creep/landslides.

Figure 12.2. Failure mechanism on the upstream (river side) slope during a rapid retreat of flood waters

VISUAL INSPECTION OF DIKES: INITIAL INSPECTION AND ROUTINE SURVEILLANCE

3.1 Justification for, principles and frequency of visual inspections

Many of the structural anomalies that can affect a dike and its appurtenant works are revealed by indicators on the surface – movement or unevenness of the ground, soil or gully erosion, areas of unusual vegetation, leaks, animal burrows, pipe outflows, cracks, displacements, etc. Visual inspection is the best way of detecting such indicators and is essential for compiling a report on a dike's initial condition (initial inspection) and for subsequent surveillance (routine surveillance).

The general principle of surveillance carried out on behalf of the dike operator consists of covering the entire length of the dike on foot and recording all visual information about existing or presumed anomalies affecting any of its constituent parts. The standard surveillance dossier in Appendix 3 contains a method of proceeding as well as standard documents (anomalies record forms).

For dikes bordered by rivers, and whenever necessary, surveillance should be periodically supplemented with:
– An inspection by boat (where the toe of the slope is steep, inaccessible and/or wooded).
– An underwater inspection (when stone facing or a protective feature at the toe of the slope extends below the low-water line).

The frequency of inspections should be geared to both the extent and value of the protected land and infrastructures and the degree of hydraulic loading to which the dike is subject. The following recommendations can therefore be made:
– At least one inspection a year on foot for dikes not subject to regular flooding.
– Two inspections a year for dikes regularly subject to minor floods and for dikes protecting valuable land and infrastructures.
– An annual inspection by boat if necessary.
– An inspection after every major flood.

These intervals may appear to be long[1], but they are minimum recommendations to be adapted to each particular situation. However, we believe it is preferable to carry out inspections less frequently, but in a thorough manner. Care should be taken not to fall into a routine that becomes synonymous with loss of efficiency.

3.2 Operating conditions and procedure

Tours of inspection should be carried out after ground cover and bushes have been carefully cleared and, if possible, during dormancy of the vegetation (autumn and winter) so as to benefit from the best possible visibility.

Operations in the field are carried out by a team of two (or three) people who are familiar with the structures (dikereeves, if any, or technicians employed by the manager). It is important for inspections to be carried out by at least two people, both to ensure the completeness and relevance of data collected and for safety reasons.

Prior to inspection, the following should be made ready:
– Dike plans, maps and cross-sections to help with identification and the recording of observations. Maps should ideally be on a scale of 1:500.
– Detailed plans of gated structures (sluices, flap gates & spillways).
– Documents containing observations from the previous inspection(s), to compare changes in deterioration.

Field technicians must be dressed appropriately (boots or waders, life jackets for inspections by boat or on steep river-side slopes, etc.). It is a good idea to have a camera to take photos of anomalies to be able to objectively compare observations on successive dates. It is also necessary to have something with which to take notes; a portable dictaphone is very useful for this.

---

1. In far too many cases, inspection intervals are even longer than these recommendations.

Information can be recorded on an anomalies record form (a blank copy is given in Appendix 3) that can be modified to suit the particular structures.

When this device can be used (in the absence of dense foliage), a GPS receiver may prove to be very useful for localisation in the field.

3.3 Fill dikes

3.3.1 Features to inspect and information to log

If, as would be ideal, a detailed topographical map is available, the information shown on it should first be checked and completed, which requires keeping track of one's position on the existing map as the inspection progresses.

Cross-sections are drawn of areas where particular features are not visible or are poorly-marked on the map (e.g. house or constructions built near, on or in the levee). Water levels on the day of the inspection should also be noted (height of the river and stretches of water).

It is advisable to inspect anomalies and deterioration by working methodically along the dike: a field method for doing so is suggested in the standard dossier for dike surveillance given in Appendix 3. Features to inspect are listed in Tables 1 and 2, using a double entry of possible failure mechanisms and by examining three different parts of the structure in the case of fill dikes. Table 1 (page 38) is intended for initial inspections and Table 2 (page 40) for routine surveillance. These two tables have many points in common since the same indicators are sought in both operations (carried out in "dry" conditions). Obviously, routine surveillance also aims to identify changes to damaged areas, which implies being in possession of reports from previous inspections.

When examining particular features, special attention should be paid to houses, buildings, tunnel and pipe outflows and observation holes located near, or built into, the body of the dike. Low points in the crest, often gated and associated with circulation through the dike, should also be examined. Particular features should be described in detail and accurately marked on the map (elevation and cross-section) when the available topographical map shows them only partially, if at all.

If the dike is fitted with readable monitoring instruments such as piezometers, the measurements should be recorded (possibly in two stages if prior maintenance work is needed). Piezometer readings may be worth taking more frequently than for routine inspections; for example, tracking of seasonal fluctuations in water table levels, which may be necessary for an in-depth dike diagnosis, will require readings to be taken at least once every three months.

Lastly, any local residents encountered at the time of inspections can be asked about the dike, its operation and any recent maintenance work that may have been carried out. Their observations can be noted in the comments sections of anomalies record forms.

Table 1. Initial visual inspection of fill dikes – summary of features to inspect

Failure mechanisms	Features to inspect	Initial visual inspection		
		River-side slope	Crest	Land-side slope
Overtopping	Longitudinal profile of the crest		Uneven profile, low points, collapsed areas, ruts – presence & condition of stop logs, gates, etc.	
	Height of the water, high-water marks	Signs of historical floods, water height on the day of the inspection, existence of flood-water debris/marks		
	Overspill		Existence, nature & state of facing and of a fuse plug (spillway)	Existence, nature & state of facing and of downstream dissipator (spillway)
	Freeboard feature		Existence, nature & state of freeboard feature: appearance of contact with dike body, stability	
Surface erosion/scouring	Effect of the watercourse's hydraulic load on the slope	Slope verticality, dislodgement of riparian vegetation, presence of an eroded bend	Longitudinal cracking on the crest coinciding with the eroded bend	
	Surface protection (facing)	Existence, nature & state of protective facing (stone facing, concrete facing, riprap, etc.).		Existence, nature & state of protective facing (river runoff on land side)
	Protection of the toe of the slope	Existence, nature & state of protection of toe of slope (pile or sheet pile protection, riprap, etc.).		
	Proximity & alignment of the main channel/flow characteristics	To examine: is the dike in direct contact with the main channel? Meander – concave bend. Speed & direction of current.		
	Effect of various external loads on the slope	Existence & stage of development of gullies, impact of earthworking, etc.		Existence & stage of development of gullies, earthworking impacts, etc.
Internal erosion	Vegetation	Nature, development & stability, roots & stumps, on or at the toe of the slope	Nature & development, roots and stumps	Nature & development, roots & stumps, on or at the toe of the slope
	Burrows & warrens	Size, location & density, evidence of recent activity	Size, location & density, evidence of recent activity	Size, location & density, evidence of recent activity

Failure mechanism	Features to inspect	Functional boundaries		
		River side	Crest	Land side
Internal erosion (*cont'd*)	Conduits, culverts & pipe crossings	Transverse conduit, culvert or pipe outlets (existence, characteristics), appearance of contact with embankment, non-return device	Conduit observation holes, over-dike pipes	Transverse conduit, culvert or pipe outlets (existence, characteristics), appearance of contact with embankment, sluice
	Upgrading work	Existence, nature & state of impervious shoulder, geomembrane, etc.	Existence, nature & state of impervious curtain (sheet piles, diaphragm wall, etc.)	Existence, nature & state of draining shoulder, etc.
	Particular features	Identification & characterisation – ladder, slipway, ramp, embedded building, etc.	Identification & characterisation – gate, stop log assembly, embedded building, etc.	Identification & characterisation – sump, retaining wall, embedded building, etc.
	Seepage/leaks	Sink holes	Sink holes	Evidence of seepage/leaking
General instability	Waterlogging, piezometry	Wet areas, spring – existence of piezometers & measurement of water level, if possible	Existence of piezometer & measurement of water level, if possible	Existence of piezometers at the toe of the slope, existence of wells or ditches, and measurement of water level(s) if possible
	Dike cross-section	Steepness of slope, presence, nature & state of shoulder, berm, etc.	Crest width	Steepness of slope, presence, nature & state of a draining shoulder
	Ground movement	Cracks in the ground, bulging, soil creep deformation, slides – damage (cracks, overturning) to rigid works – leaning trees	Longitudinal cracks, collapsed areas – damage (cracking, overturning) to rigid works such as roadways, parapets, walls	Cracks in the ground, bulging, soil creep deformation, slides – damage (cracks, overturning) to rigid works – leaning trees
Breach	Evidence of historical breaching	Localised modification of dike profile or nature	Localised modification of dike profile or nature. Stele! (in memory of a dike patron – e.g. Conneuil breach on the left-bank levee of the Loire upstream of Tours)	Depression, pond or marsh beyond the toe of the slope. Localised modification of dike profile or nature
	Accessibility for earthworking (& maintenance) machinery	Of no interest with respect to the risk of breaching (site not accessible for operations during flooding), but purely for routine maintenance of the lower part and toe of the slope	Existence, characteristics and practicability of roadway	Existence, characteristics and practicability of the roadway at or near the toe of the slope

Visual inspection of dikes: initial inspection and routine surveillance

3.3.2 Recording and writing up information

Details of anomalies and other relevant information are recorded in the anomalies record form (Appendix 3).

A comprehensive photographic record, suitably captioned, geographically identified and dated, is also to be compiled with:
– Photos of deterioration, with references and captions.
– Photos of the overall structure.

Table 2. Routine visual surveillance of fill dikes – summary of features to inspect

Failure mechanisms	Features to inspect	Routine visual surveillance?		
		River-side slope	Crest	Land-side slope
Overtopping	Longitudinal profile of the crest		Appearance/changes in profile unevenness – low points, collapsed areas, ruts – condition of stop logs, gates, etc.	
	Height of the water, flood-water marks	Water height on the day of the inspection, existence of recent flood-water marks		
	Overspill		State of facing and of any fuse plug (spillway)	State of facing and of downstream dissipator (spillway)
	Freeboard feature		State of the freeboard feature: appearance of contact with dike body, stability	
Surface erosion/scouring	Effect of the watercourse's hydraulic load on the slope	Slope verticality, dislodgement of riparian vegetation, appearance/changes in eroded bend	Longitudinal cracking on the crest coinciding with the eroded bend	
	Surface protection (facing)	State of protective facing (stone facing, concrete facing, riprap, etc.).		State of protective facing (river runoff on land side
	Protection of the toe of the slope	State of protection of toe of slope (pile or sheet pile protection, riprap, etc.).		
	Proximity & alignment of the main channel/ flow characteristics	To examine if the levee is close to the main channel: State of contact with main channel. Speed & direction of current.		
	Effect of various external loads on the slope.	Appearance and/or stage of development of gullies, impact of earthworking, etc.		Appearance and/or stage of development of gullies, impact of earthworking, etc.

Table 2. (cont'd)

Failure mechanism	Features to inspect	Central visual surveillance — Crest of the dike	Central visual surveillance — Upstream slope	Central visual surveillance — Downstream slope
Internal erosion	Vegetation	Nature, development & stability, roots & stumps, on or at the toe of the slope	Nature & development, roots and stumps	Nature & development, roots & stumps, on or at the toe of the slope
	Burrows & warrens	Size, location & density, evidence of recent activity	Size, location & density, evidence of recent activity	Size, location & density, evidence of recent activity
	Conduits, culverts & pipe crossings	Transverse conduit, culvert or pipe outlets, appearance of contact with embankment, state of any non-return device	Conduit observation holes, over-dike pipes	Transverse conduit, culvert or pipe outlets, appearance of contact with embankment, state of any sluice gates
	Upgrading work	State of any impervious shoulder, geomembrane, etc.	State of any impervious curtain (sheet piles, diaphragm wall, etc.)	State of any draining shoulder, etc.
	Particular features	State, configuration – ladder, slipway, ramp, embedded building, etc.	State, configuration – gate, stop log assembly, embedded building, etc.	State, configuration – sump, retaining wall, embedded building, etc.
	Seepage/leaks	Sink holes	Sink holes	Evidence of seepage/leaking
General instability	Waterlogging, piezometry	Appearance/changes in wet areas, springs. State of piezometers & measurement, if possible	State of piezometers & measurement, if possible	State of piezometers at the toe of the slope, existence of wells or ditches, measurement of water level(s) if possible
	Ground movement	Appearance/changes in cracks in the ground, bulging, soil creep deformation, slides – damage (cracks, overturning) to rigid works – leaning trees	Appearance/changes in longitudinal cracks, collapsed areas – damage (cracks, overturning) to rigid works such as roadways, parapets & walls	Appearance/changes in cracks in the ground, bulging, soil creep deformation, slides – damage (cracks, overturning) to rigid works – leaning trees
Conditions of access for maintenance	Accessibility for earthworking (& maintenance) machinery	State of roadway at toe of slope	State of roadway on crest	State of roadway at or near the toe of the slope

(*) Obtain information/reports from previous inspection

3.3.3 Output and limits of visual inspection

The output of visual inspections depends directly on the conditions in which they are carried out – that is, whether or not detailed maps and plans are available and, particularly, the state of the vegetation. It is not difficult to imagine the difference in the difficulty, speed and visibility of inspections between a well-maintained dike, where the ground cover is kept short, and a neglected dike covered in (frequently prickly...!) scrub growth such that progress can only be made with a machete(!), and when the only documentation available is a 1:25,000 map.

Obviously, the first visit (initial inspection) takes longer than subsequent ones (routine surveillance), when it is 'simply' a matter of updating information.

We venture to give a few figures for the case of a well-maintained dike, for which accurate plans are available, not including the time needed to transcribe information back in the office:
– Initial inspection: 1 to 2 km a day per field team.
– Routine surveillance: 3 to 5 km a day per field team.

In the case of a poorly-maintained dike, this output can be halved and, in extreme cases, reduced even more.

Visual inspections are limited in that they do not provide information about damage (in principle associated with phenomena occurring underground and/or with the behaviour of watercourses in the vicinity of dikes during flooding) that does not produce (or has not yet produced) signs at the surface. This is particularly so for dikes in non-flood conditions (e.g. areas of greater permeability in the dike's body or foundations, piping that has yet to find an outlet, forces exerted by riverbank currents, etc.) and those where such signs have disappeared (e.g. buried constructions or structures, remodelled areas of ground movement or unevenness, former site of overtopping, etc.).

In this respect, the risk of an inspection not being exhaustive is greater the longer ago the structure was put under load (major flood), which is why it is a good idea, whenever possible, to carry out inspections during and/or after flooding (sections 4.2 and 4.3 respectively) as well as in 'dry' conditions.

Even so, initial visual inspection is fundamental to dike study-diagnosis. It should be carried out prior to geotechnical exploration, the implementation of which it will help determine. Subsequent routine inspections, which are less time-consuming, make it possible to update evaluations of existing conditions.

3.4 Masonry and concrete walls

The principal types of damage to be sought in masonry or mass concrete structures can be divided into three groups: structural damage, scouring and local deterioration. Table 3 (page 43) contains a summary of features to examine as part of a visual inspection.

3.4.1 Structural damage

Structural damage takes the form of cracks, which usually affect the entire structure across its width. They are due to differential settlements of foundations or to active earth pressure held back by the wall. They appear in three forms (Fig. 13) – opening, horizontal throw (relative forward-back displacement) and sliding (relative vertical displacement).

In principle, cracks characterised by throw are a sign of active earth pressure, whereas cracks characterised by sliding or opening are more likely to be linked to the differential settlement of foundations.

Table 3. Visual surveillance of masonry and concrete dikes, spillways and particular features – summary of features to inspect

	Features to inspect			
Structural movement	Settlement		Cracks characterised by opening or sliding, uneven profile, low points, collapsed areas	
	Land thrust	Transverse cracks with throw	Transverse cracks with throw	Transverse cracks with throw
	Concrete shrinkage	Transverse cracks without throw or sliding	Transverse cracks without throw or sliding	Transverse cracks without throw or sliding
Scouring/erosion	Effects of the watercourse's hydraulic loads	Dislodgement of the base of the wall, undercutting, presence of an eroded bend	Cracks characterised by opening or sliding, uneven profile, low points, collapsed areas	
	Effects of overtopping on spillways		Stones swept away from the overflow sill	Stones swept away from the discharge chute or invert, excavation of scour holes downstream of the invert, undercutting of the invert
	Protection of the toe of the slope	Existence, nature & state of protection at toe of slope (pile or sheet pile protection, riprap, etc.).		
	Proximity & alignment of main channel/flow characteristics	To examine: is the dike in direct contact with the main channel? Meanders – concave bends. Speed & direction of current		
Localised deterioration	Ageing of stones	Stones that are cracked, split, shattered by freezing or missing	Stones that are cracked, split, shattered by freezing or missing	Stones that are cracked, split, shattered by freezing or missing
	Ageing of masonry joints	Damaged joints, cracked, porous, crumbling mortar	Damaged joints, cracked, porous, crumbling mortar	Damaged joints, cracked, porous, crumbling mortar
	Vegetation	Nature & development of plant life growing in masonry joints	Nature & development of plant life growing in masonry joints	Nature & development of plant life growing in masonry joints
	Particular features, conduits, culverts & pipe crossings, embedded buildings	Location & characterisation. Transverse conduit, culvert or pipe outlets (existence, characteristics), appearance of contact with concrete or masonry, non-return device	Location & characterisation. Conduit observation holes, over-dike pipes, state of stop log grooves.	Location & characterisation. Transverse conduit, culvert or pipe outlets (existence, characteristics), appearance of contact with concrete or masonry, sluice
	Repairs	Existence, type of repairs (re-jointing, replacement of stones, etc.)	Existence, type of repairs (re-jointing, replacement of stones, etc.)	Existence, type of repairs (re-jointing, replacement of stones, etc.)
Damage to spillway erodible ridges	Settlement & erosion		Evenness of longitudinal profile, low points, gullies created by rain, damage caused by animals or vehicles	

Concrete walls generally have construction joints placed at regular intervals, which allow for thermal contraction as the concrete sets, as well as seasonal distortion due to changes in temperature. Without construction joints, shrinkage cracks often appear at regular intervals with no differential displacement perpendicular to the wall. Such cracks are not especially serious. However, construction joints – or shrinkage cracks that appear in the absence of joints – can be the source of movement other than a simple opening, thus materialising the presence of structural problems.

If such cracks or signs of displacement are noticed on a masonry or concrete wall, a civil engineering specialist should be called in to make a more accurate diagnosis so as to find the cause, evaluate the implications and recommend any necessary corrective measures. The specialist may suggest monitoring the crack using a simple fissurometer (allowing measurement in one direction only) or, preferably, a jointmeter, which measures relative displacement in three orthogonal directions (Fig. 13).

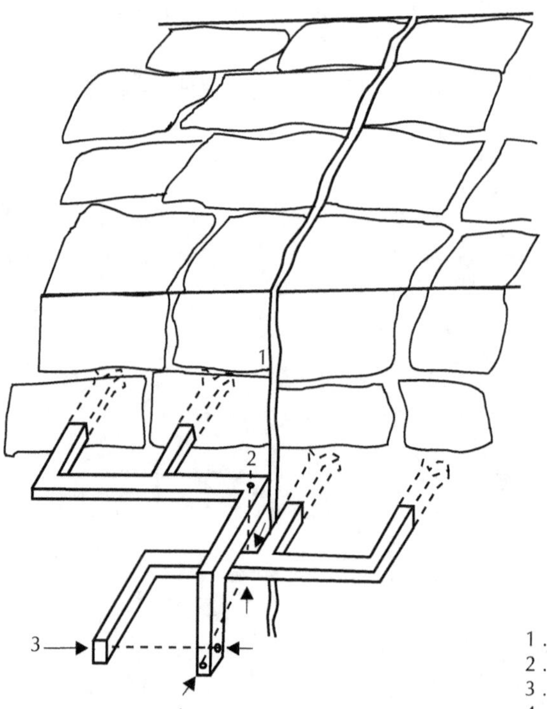

Figure 13 Movement around a crack and monitoring with a jointmeter

A three-dimensional jointmeter consists of two mating elbow-shaped brackets that are embedded into and straddle the crack. The cross-section of the metal brackets must be at least 2 cm x 2 cm. Displacement readings are obtained with a calliper gauge with an accuracy of 1/100 mm. The overall accuracy of the reading depends very much on the care taken to embed the apparatus and on the rigidity of the metal brackets. The two brackets must be properly anchored into cut stones that are firmly attached to each section of wall on either side of the crack (avoid stones that have come loose).

3.4.2 Scouring

Masonry dikes are often built where space is restricted and are therefore frequently to be found where dike-limited floodplains are narrowest. More often than not, they are in direct contact with the main channel, which makes them particularly susceptible to scouring.

Particular attention should therefore be paid to examining the base of the wall on the river side, which should logically be done when the water level is low, by boat if need be. An underwater inspection may also be considered necessary. It is advisable to periodically conduct a survey of the river bed at the base of the wall using, for example, sounding rods, sounding lines (weighted, graduated lines) or graduated levelling staffs. A sonar is useful when the whole length of the wall can be covered by boat, which may be easier when the river is at mean water level. In high-risk sectors, bed surveying (using sounding rods or sonar) should be done after each major flood.

3.4.3 Localised deterioration

This involves deterioration due to the ageing of stones or the masonry binding agent. Stones may suffer from the effects of freezing and thawing and, less frequently, mechanical erosion or weathering (e.g. freestone in the Loire).

Mortar may be affected by physico-chemical deterioration: depending on its quality at the outset, cement is subject to chemical weathering, which weakens the mortar, making it porous and, therefore, susceptible to freezing – thawing cycles, the development of vegetation and water erosion. Soil particles may also be deposited on masonry joints, providing an environment that is conducive to the germination of seeds and the establishment of plants, the roots of which can eventually break up masonry joints and even cause more widespread damage to structures themselves if the roots occupy the whole thickness of the joint.

3.5 Spillways

Damage to concrete and masonry spillways is similar to that described above for concrete walls – structural cracks and ageing of concrete or masonry.

Deterioration of spillways protected by a masonry facing is described in section 3.4.3.

Spillways may also be damaged by water erosion, such as stones being washed away from the sill, discharge chute or downstream invert. The area downstream of the invert may also be affected in the form of scouring, which may even extend to undercutting the invert.

These phenomena obviously occur only when there have been floods the magnitude of which has activated the spillway. Special attention should therefore be paid when examining spillways after flooding. All such damage occurring during flooding is serious and should be repaired rapidly, or at least before the next flood. If not, deterioration will get considerably worse and may lead to total failure of the spillway.

Points to be examined during visual inspection are summed up in Table 3.

3.6 Particular structures and features

As mentioned in section 1.3.4, particular features may vary widely in nature – stop log structures, ramps giving access to the river, aqueducts, tunnels, conduits, culverts, pipes, construction in the dike body, etc. We will not, therefore, attempt to list the damage that can affect these structures.

In reality, the first job to be done during surveillance of these structures is to describe them as fully as possible and mark them accurately on maps and plans. Inspections provide an opportunity to identify any damage (see Table 3) and list particular features that have yet to be catalogued. For this category, photography would be an excellent way in which to monitor damage.

4 SURVEILLANCE DURING FLOODING

4.1 Importance of preparation (flood alert plan and/or action plan)

Since, by their very nature, flood situations are foreseeable only a (very) short time in advance, good practice demands that detailed preparations be made for such an eventuality, especially in terms of:
– Organisation of flood warning.
– Installation of stop logs, gate manoeuvring, verification of flap gates and valves.
– Surveillance of structures during flooding.
– Where appropriate, procedure for evacuating people threatened by a rising water level or dike failure.

This handbook does not seek to cover the details of flood alert plans, which are specific to each particular locality. However, a few general recommendations can be made here, while section 4.2 details procedure for flood surveillance.

In France, **flood warning** is not officially organised for all rivers. When such a system does exist, the flood alert plan should indicate the ways of communicating information from the flood warning service to the dike operator services. Indications about the expected time of arrival of flood waters are also invaluable.

If there is no flood forecasting service for the river, the shortest notification chain should be identified, from reports issued by the national meteorological service (e.g. Météo France) and limnigraphic stations to the operators of dikes the failure of which may threaten "public safety" (see Appendix 5, paragraph 6).

The actions taken to operate shut-off devices must be meticulously prepared, starting with a complete list of structures, their characteristics and location on a map, places where stop logs are kept, etc. Practice sessions in handling and installing this equipment should be organised periodically out of flood periods, and specific verifications made when flood warnings are issued. Regular maintenance of such equipment ensures that it will be operational when needed.

In terms of organising surveillance during flooding, sectors and specific points in the dike that require priority inspection need to be identified based on the operator's knowledge of the condition of the dike (obtained from initial and subsequent routine inspections) and of the interests and assets it protects. A summary report of the work to be carried out should be compiled for each sector or point to be inspected (KM identification, notes and KM coordinates of specific points to be examined during linear inspection, general checklist of anomalies to look for, frequency of the operation if it is considered necessary to repeat it a number of times during flooding, etc.).

The summary report should also mention any documents and equipment the field team needs to take along (making sure that the service has enough copies available).

It is possible to define different degrees of flood alert, depending on the seriousness of the flood forecast. This particularly applies to large rivers for which warning systems and flood propagation times make such a gradation of degrees possible in the organisation of their flood alert plans.

The safety of those who have to work in the field should be a permanent concern in the drafting of a flood alert plan. Special safety and communications equipment should be provided, notably life jackets and walkie-talkies[1].

4.2 Visual surveillance during flooding

4.2.1 Justification for and principle of the method

The general purpose of inspecting dikes during flooding is to identify, record and assess anomalies or suspected anomalies associated more or less directly with the

---

1. The use of mobile phones is not necessarily appropriate in crisis situations when networks are likely to be congested.

"loaded" state of a dike, revealing the structure's weak points (in addition to those identified during 'dry' inspections) and/or areas that may point to potential failure in the future. These anomalies may be the result of the external hydraulic or mechanical stresses to which the dike is subject (hydraulic head, overtopping, bank-side current, waves, hydraulic jumps and turbulence) or to internal mechanisms initiated by a water-level related effect or phenomenon (circulation of water through or under the body of the dike, saturation levels, hydraulic currents or pore pressure).

The principle of the method involves walking along a length of dike that is under load during a flood. It may therefore be that this inspection is carried out in a crisis situation (with a flood warning in place, or even with disaster relief teams deployed). Compared with 'dry' inspections, the advantage of surveillance during floods is that useful information can be gathered on anomalies or changes in anomalies linked to the nature of the soil within the dike (e.g. areas of greater permeability in the body of the dike, signs of internal erosion, etc.) and/or to the behaviour of the river in flood in front of the levee (e.g. overtopping, bank erosion, etc.). The problem is that such signs may be noticed only a very short time before the sudden failure of the structure.

Apart from the nature of those indicators of anomalies that are to be given particular attention, visual surveillance during flooding differs from 'dry' inspections in several important ways:
– It takes two forms, neither of which is exclusive:
  • *Linear* inspection (possibly repeated during the flood) of a *pre-determined sector* with the aim of verifying the dike's critical functioning and adding to existing information about the dike and its watertightness.
  • *Intermittent* (and possibly repeated) inspection of a *limited, strictly defined area of dike* where damage (leaks, overtopping, etc.) is reported during a flood by witnesses or where anomalies have previously been suspected (e.g. particular features).
– The day and time of year of the inspection are dictated by events, making preparation time more or less short. If the slopes and/or edges of embankments are poorly maintained (ground cover), there will hardly be time to clear vegetation beforehand (hence the importance of keeping works well-maintained to ensure optimum visibility at all times).
– Observations made during in-flood surveillance can be integrated into an ongoing crisis management process and help determine ways to evacuate areas where populations are at risk or conservation repairs are to be done by manual or mechanical means (e.g. sealing of pipes, reinforcements, plugging of breaches, etc.). A maximum length of dike therefore needs to be covered in a minimum amount of time. Attention needs to focus on the most crucial points and field technicians need to be equipped with rapid means of communication.
– Developments being monitored may evolve very rapidly and the collection of information needs to be finely timed (to the minute or, at the very most, quarter of an hour).
– Field team members are potentially exposed to physical risks, and measures should be taken to ensure their safety.

All these considerations dictate that the practical aspects of conducting in-flood inspections should be established outside of crisis periods (flood alert plan: see section 4.1 above). If possible, practical exercises should also be conducted.

4.2.2 Operating conditions and procedure

On principle, in-flood surveillance can be carried out no matter what the type of terrain and access, but its *effectiveness and output depend directly on the state of the dike's ground cover* at the time of the operation. Since the timing of such inspections cannot be known in advance, the only way to ensure good visibility is to have the dike and its immediate surroundings properly maintained at all times.

In principle, it is the dike operator's service that prepares and organises visual inspections during flooding. However, since inspections are likely to take place in a process of crisis, they may involve other people, on both the decision-making and operational levels, which can lead to communication and coordination problems. Logically then, most preparations for this type of inspection should be made outside of crisis periods and written into the *flood alert plan*.

The flood alert plan is drawn up out of a fundamental concern for the safety of those working in the field and for the efficiency of communications and decision-making. It clearly identifies the teams of people to be mobilised and, for each team, the segment of dike to be examined. If need be, the plan also makes provision for helicopter assistance in the visual surveillance of dikes during flooding: evacuation/rescue of team members, supply of equipment or materials for conservation repairs, etc.

The field team consists of two people, at least one of whom should be relatively conversant with civil engineering and soil mechanics. Working in pairs is advisable and even essential for carrying small items of equipment, for conducting rapid surveys in good conditions and for the safety of those involved. If one of them is familiar with soil mechanics, it ensures the relevance of observations and the correct assessment of short-term risks in terms of the safety of the structure and, therefore, of the team members themselves. It is also a good idea if one of them is the general manager of the segment of dike in question (responsible for routine inspections and checking maintenance work).

The length of the segment assigned to each team will depend on three conditions:
– How safe the segment is, which is determined by the diagnostic work that has been carried out in "dry conditions"; a segment that has previously shown signs of anomaly or weakness should be monitored more closely.
– Conditions for carrying out inspections – ways of getting about on the dike and state of the vegetation.
– The degree of vulnerability of areas protected by dikes, given the hydraulic risk and the importance of nearby interests and assets (housing, infrastructures, public services, valuable crops, etc.).

As a general rule, the length of dike assigned to a given team should not be more than about twenty kilometres.

4.2.3 Features to inspect and information to log

Particular points to look for are outlined in Table 4, on the basis of a double entry of possible failure mechanisms and the three parts of the structure to be examined in the case of a fill dike.

Table 4. Visual surveillance of fill dikes during flooding – summary of features to inspect

Failure mechanism	Features to inspect	Parts of the structure that can be examined		
		River side slope	Crest	Land side slope
Overtopping	Longitudinal profile of the crest		Verification that stop logs are in place – behaviour under load (stability, seal, etc.)	
	Height of the water, flood-water marks	Measurement of the height of the watercourse (at least in relation to the crest). Location of high-water marks	Signs and location of recent overtopping – debris, marks, flattened grass, etc.)	Signs and location of recent overtopping – debris, marks, flattened grass, etc.)
	Overspill		If overtopping apparent => **Alert!** Height of overtopping – resistance of crest to gully formation. Spillway – operational or not? State of fuse plug, spillway behaviour	If overtopping apparent => **Alert!** Height of overtopping – resistance of crest to gully formation. Spillway – operational or not? Behaviour of discharge chute & energy dissipator. Extent of flooding on land side (visual)
	Freeboard feature		Behaviour of freeboard feature under load – appearance of contact with dike body, seal, stability	
Surface erosion/scouring	Effect of the watercourse's hydraulic loads on the slope	Beginnings or further development of eroded bend. Destabilisation of trees, cracks on top of slope	Longitudinal cracking, collapsed areas on crest, damage to rigid structures in the area under attack on the river side. If crest erosion apparent => **Alert!**	Water from river on land side & any impact at the toe of or on the dike slope
	Surface protection (facing)	Protective facing's resistance to erosion, signs of movement		Resistance of protective facing on land side, if it exists
	Protection at toe of slope	In principle, cannot be examined		
	Proximity & alignment of main channel/flow characteristics	Direction & speed of bank-side current. Existence & size of waves, eddies, jumps, whirlpools, vortexes, etc.		

Table 4. (*cont'd*)

Failure mechanisms	Features to inspect	Visual surveillance during flooding		
		River-side slope	Crest	Land-side slope
Internal erosion	Vegetation			▌Check for signs of seepage
	Burrows & warrens	Location & examination of large burrows & warrens	Location & examination of large burrows & warrens	Location of large burrows & warrens. Check for signs of leaking
	Conduits, culverts & pipe crossings			Check for signs of seepage/leaks
	Upgrading work			Check for signs of seepage/leaks
	Particular features			Check for signs of seepage/leaks
	Seepage/leaks			Leaks, seepage, rivulets, wet or waterlogged areas on the slope or its facing, at the base of stumps, at openings to burrows, culverts, conduits, pipes & land-side consolidation drains, on embedded buildings or other particular features. Sand-boils or reappearances beyond the toe of the slope, in ditches, channels, depressions, sumps, wells, etc.
	Beginnings of piping	▌Sink holes, unusual cavitation, whirlpools, vortexes	▌Sink holes, unusual cavitation	Check for turbidity of all water flows identified
General instability	Waterlogging, piezometry		▌Check the bearing capacity of the soil. Piezometer readings	▌Check the bearing capacity of the soil. Piezometer readings, measurement of water level in sumps, wells, etc.
	Ground movement	Signs of ground movement – cracks, mounds, bulging, sliding) at retreating water level stage	Longitudinal cracking, collapsed areas – damage (cracking, overturning) to rigid structures such as roadways, parapets ▌& walls	Cracks in the ground, mounds, bulging, soil creep deformation, slides – damage (cracking, overturning) to rigid structures ▌– leaning trees
Breach	Accessibility for earthworking machinery		Verification that crest roadway is passable	Verification that toe roadway is passable

▌Signs to look for in particular

If the dike is equipped with readable monitoring instruments such as piezometers, measurements should be taken when possible, at least from instruments that can be reached without danger.

Additionally, photographs of the most serious anomalies are useful if taken with an instant camera (digital camera), so as to obtain images that can be used quickly.

4.2.4 Recording and writing up information

It would be difficult to insist on filling in anomalies record forms in the field, not least during crisis situations (major flooding). For the sake of rapidity, a notebook can be used to record information simply, mentioning references to kilometre markers (or GPS waypoints if such a device can be used), a basic transverse datum point (e.g. lower – middle – upper land-side slope), a short description possibly accompanied by a sketch, the references of any photos taken and the date and time. A dictaphone is useful for quickly recording all the information above, especially if the operation takes place during the night or in the rain.

If a detailed topographical map of the dike is available (1:500 or 1:1000), observations can be marked to scale on a copy of the cartographic document, with a number or code referring to a description in the notebook or on the dictaphone recording. It is also a good idea to mark the angle from which any photos have been taken.

In any event, once the crisis situation is over, field notes and/or voice recordings need to be used by the operator to supplement information about the dike. In this respect, it is recommended to carry out a post-flood inspection (see section 4.3) for the purpose of validating observations, and their references to kilometre markers, made during flooding and assessing any flood-induced changes to anomalies. Data collected during and after flooding can then be logged later in the form of record forms for future reference, including IT methods (statistics, anomalies data analysis, etc.).

4.2.5 To sum up

Inspections carried out during flooding offer two ways to learn more about dikes:
– Data can be collected about the behaviour of dikes under hydraulic load, when they are habitually 'dry', *hence the advantage of doing an in-flood visual inspection even if the hydraulic head is only partial (medium-intensity flood)*.
– In high-risk areas (high failure probability and great vulnerability), they make it possible to evaluate and monitor dike safety in crisis situations (major flooding).

Nevertheless, operations to be conducted in order to carry out an in-flood diagnosis should be decided **beforehand** in a detailed flood alert plan, in which all practical aspects must be specified:
– The personnel that can be assigned to each dike sector.
– Distribution of jobs and previous training.
– List of points to be given particular attention.
– Safety instructions and equipment geared to the risks incurred by team members.
– If need be, any helicopter transport that can be mobilised.

Finally, the inspection is only of value if subsequently written up (with checklists and record forms, if possible) and accompanied by photos, drawings, etc.

4.3 Post-flood visual surveillance

4.3.1 Justification for and principle of the method

The general purpose of post-flood inspection is to identify, record and assess anomalies or suspected anomalies associated more or less directly with the "loaded" conditions to which a dike has just been subjected. It *is a kind of 'special routine inspection' conducted after the flood* that reveals the structure's weak points (in addition to those detected during 'dry' inspections) and/or, if carried out after one or more in-flood inspections, to validate, verify and add to information collected at that time. Besides this, it can be used as a basis for establishing a programme of urgent work designed to repair the worst damage the dike, or its spillways, has suffered during the flood.

The principle of the method involves walking along a length of dike that has recently been under load after a river has been in spate. This inspection may therefore come after one or more in-flood visits to all or part of the length of dike in question, in which case, information collected previously can be verified and supplemented. Compared with 'dry' inspections, the advantage of post-flood surveillance is that useful information can be gathered on anomalies or changes in anomalies linked to the nature of the soil within the dike (e.g. areas of greater permeability in the body of the dike, signs of internal erosion, etc.) and/or to the behaviour of the river in front of the dike during recent flooding (e.g. overtopping, bank erosion, etc.).

4.3.2 Operating conditions and procedure

On principle, post-flood surveillance can be carried out no matter what the type of terrain and access, but *its effectiveness and output depend directly on the state of the dike's ground cover* at the time of the operation. Since the timing of these inspections cannot be known in advance and there is only a short period of time (several days) in which to carry them out, the only way to ensure good visibility is to have the dike and its immediate surroundings properly maintained at all times.

In principle, it is the dike operator's service that prepares, organises and carries out post-flood visual inspections. However, since it is vital for an inspection to be made as soon as possible after a flood, the operator may decide to call in people from other services or even a specialist service provider. If not carried out by the operator's staff, the latter should at least be involved in the preliminary and reporting stages.

In preparation for the visit, all relevant topographical documents (updated at the time of the last routine inspection or put together subsequently) should be collated to serve as an aid in the field. All record forms and other documents from all previous visits (routine and in-flood) are then analysed to identify the special or evolutive points that will need to be examined at the time of the next visit.

The team in the field consists of two or three people, at least one of whom should be relatively versed in civil engineering/soil mechanics. Working in pairs is advisable and even essential for carrying small items of equipment, for conducting rapid surveys in good conditions and for the safety of those involved.

4.3.3 Features to inspect and information to log

Anomalies, the evidence of which is to be sought in particular, may be the result of external hydraulic or mechanical stress to which the dike has been subject (hydraulic head, overtopping, bank-side current, waves) or of internal mechanisms initiated by a water-level related effect or phenomenon (circulation of water through or under the body of the dike, saturation levels, hydraulic currents or pore pressure).

Particular points to look for are outlined in Table 5, on the basis of a double entry of possible failure mechanisms and the three parts of the structure to be examined in the case of a fill dike.

As mentioned in section 3.6, particular attention should be paid to examining spillways, especially if they were activated during the flood peak. The priority objective is to locate all traces of erosion and scouring.

If the dike is fitted with readable monitoring instruments such as piezometers, it is appropriate to take the measurements (possibly in two stages if they first need maintenance work – e.g. cleaning of piezometers when the head has been under water during flooding).

Lastly, any local residents encountered at the time of the inspection should be asked about how the dike behaved during the flood. Their observations can be noted in the comments sections of anomalies record forms.

4.3.4 Recording and writing up information

Post-flood inspectors use standard anomalies record forms supplied by the dike operator. This may be the same as the form used for 'dry' inspections (Appendix 3). Recording information on a dictaphone speeds up the process. As an example, Appendix 4 outlines the methods used during post-flood inspection of the River Agly dikes following the events of November 1999.

The anomalies identified in each part of the structure are located, numbered and entered directly onto a copy of the 1:500 topographical map (if there is one), using a standard reference system. The numbers correspond to successive lines on the anomalies record form, which contains detailed notes and the main items of information in code form. If no detailed topographical map is available, work is carried out solely on the basis of cross-sections identified by kilometre markers (or by GPS waypoints if such a device can be used).

New cross-sections are drawn of places where the flood has caused substantial changes (e.g. slope erosion or landslide, main channel of the river getting closer, etc.). These cross-sections are drawn on the back of record forms and given a location in relation to the datum kilometre marker.

Table 5. Post-flood visual surveillance of fill dikes – summary of features to inspect

Failure mechanisms	Features to inspect	Post-flood visual surveillance		
		River-side slope	Crest	Land-side slope
Overtopping	Longitudinal profile of the crest		Malfunction of crest wall gates	
	Height of the water, flood-water marks	Measurement of the height of the watercourse. Examination of high-water marks	Signs & location of overtopping during flooding – debris, marks, flattened grass	Signs & location of overtopping during flooding – debris, marks in relation to flooding on land side
	Overspill		Overtopping apparent – size of overtopped area(s), state of crest, roadway & verges	Overtopping apparent size of overtopped area(s), state of slope and its toe, extent of scouring
			Spillway – did it activate or not? State of fuse plug (washed away or not?). State of spillway (invert and guide walls)	Spillway – did it activate or not? State of spillway chute and energy dissipator
	Freeboard feature		Has it been under hydraulic load or not? Appearance of contact with body of dike, stability	
Surface erosion/scouring	Effect of the watercourse's hydraulic loads on the slope	Meticulous diagnosis of the state of the slope & banks (if close to dike). Location & extent of eroded bends and/or soil creep deformations and slides. Appearance of vegetation (bank & slope), obstructions/log jamming	Longitudinal cracking, collapsed areas on crest, damage to rigid structures in the area under attack on the river side. Crest erosion – size of eroded area	State of slope and its toe vis-à-vis possible impact of water flows or inundation on land side
	Surface protection (facing)	State of protective facing – undercutting, cracking, signs of movement, functioning after drying (water flowing out through weep holes or joints)		State of protective facing on land side, if it exists
	Protection at toe of slope	State of protection at toe of slope (if visible) – undercutting, cracking, signs of movement, functioning after drying		
	Proximity & alignment of main channel/nature of flow	Modification of main channel alignment, alluvial deposits, meandering, new flow characteristics		
Internal erosion	Vegetation	Search for cavitation around stumps		Check for signs of seepage/leaks around stumps
	Burrows & warrens	Location & examination of large burrows & warrens	Location & examination of large burrows & warrens	Location of large burrows & warrens – check for signs of seepage/leaks

	Features to inspect			
Internal erosion (*cont'd*)	Conduits, culverts & pipe crossings	Search for cavitation around inlets		▌Check for signs of seepage/leaks
	Upgrading work	▌State, behaviour after drying		Check for signs of seepage/leaks around drain outlets
	Particular features	Search for cavitation on surfaces in contact with embankment		Check for signs of seepage/leaks
	Seepage/leaks			Rivulets, residual leaks, seepage, wet or waterlogged areas on the slope or its facing, at the base of stumps, at openings to burrows, culverts, conduits, pipes, land-side drains, embedded buildings or other particular features. Sand-boils and persistent reappearances beyond the toe of the slope, in ditches, channels, depressions, sumps, wells, etc.
	Beginnings of piping	▌Sink holes, unusual cavitation	▌Collapsed areas	Turbidity of residual flow water. If piping spotted: location & size of downstream orifice
General instability	Waterlogging, piezometry	▌Check the bearing capacity of the soil. Piezometer readings if functioning	▌Check the bearing capacity of the soil. Piezometer readings	▌Check the bearing capacity of the soil. Piezometer readings, measurement of water levels in sumps, wells, etc.
	Ground movement	Meticulous search for fresh signs of ground movement – cracking, bulging soil creep deformation, slides – damage (cracking, overturning) to rigid structures – leaning trees	Longitudinal cracking, collapsed areas – damage (cracking, overturning) to rigid structures such as roadways, parapets, walls, etc., notably towards the two sides of the crest	Cracks in the ground, bulging, soil creep deformation, slides – damage (cracking, overturning) to rigid structures, leaning trees
Breach	If a breach is found	▌Meticulous site diagnosis – localisation, measurements, geological cross-sections, interviews with local residents, search for causes (old conduits, pipes, tree roots, etc.), photographic reporting, etc.	▌Meticulous site diagnosis – localisation, measurements, geological cross-sections, interviews with local residents, search for causes (old conduits, pipes, tree roots, etc.), photographic reporting, etc.	▌Meticulous site diagnosis – localisation, measurements, geological cross-sections, interviews with local residents, search for causes (old conduits, pipes, tree roots, etc.), photographic reporting, etc.
	Accessibility for earthworking machinery	Possibility of access on the river side (for emergency repairs to protect the slope and/or bank)	How passable is the crest roadway?	How passable is the roadway at the toe of the slope?

▌ Signs to look for in particular

A full photographic portfolio is also to be compiled, with a key and dates:
– Photos of anomalies, identified by an anomaly reference number.
– Photos of the whole area.

Once back in the office, this information is transcribed (or written down in the case of dictaphones) and archived.

4.3.5 Anticipated output

The global output of post-flood inspections is predictably inferior to that of routine inspections since they are likely to take place at unfavourable moments (vegetation in full growth) and in places where the number of indicators to be noted is liable to be greater.

In the field, a trained three-person team should be able to cover 3 to 5 km a day. To this must be added the time it takes (probably equivalent) to transcribe findings once back in the office, something that can be postponed (but not forgotten!) in order to inspect the entire length of the dike as quickly as possible after flooding.

Whatever the circumstances, output will depend on the state of the dike and the quality of the maps and plans available, the best output being obtained when a dike is neat and tidy (ground cover cleared and cut) and with a 1:500 or 1:1000 map.

4.3.6 To sum up

Post-flood visual inspection is a very effective method for locating visual damage caused by recent loads on a dike and, therefore, for tracking down evidence of non-visible malfunctions before floods occur. It also means an "up-to-the-minute" assessment can be made of any damage that may have been caused by flooding, with a view to scheduling any necessary emergency repairs.

It should be carried out as soon as possible after flooding, when the indicators are still fresh (wet areas, high-water marks, erosion, ground movement, etc.) and before they fade or disappear. Its results, like its output, depend on the state of dike maintenance.

Notes and findings should culminate in the drafting of record forms, supplemented by photos and drawings.

5 DIKE MAINTENANCE

5.1 General principles and resources

5.1.1 Principles of maintenance

As mentioned in the foreword, in France, owners are wholly responsible for the safety of their structures and, as such, should take care of their maintenance. If the owner decides to delegate this mission to an operator, an agreement or contact should be signed that specifies the duration, exact scope and detailed content of the corresponding mission.

Regular maintenance of a high standard ensures:
- That the safety of structures is kept at a satisfactory level.
- The early detection of the beginnings of anomalies, which can then be repaired immediately at a relatively low cost, thereby preventing more substantial damage the consequences of which may be serious and prejudicial.

Dike maintenance is based on the following:
- *Visual inspection of structures – both routine and post-flood* (see Chapter 4), the latter being essential for taking stock of damage that occurs during flooding, especially on the river-side slope.
- *Keeping ground cover under control* on the dike itself and, if need be, its immediate surroundings.
- *Minimising damage caused by burrowing animals.*
- *Maintaining* parts of the structure and linear protection that contain *stone masonry, gabions, metal parts*, etc.

5.1.2 Service track

When there is no road on the dike crest, we strongly recommend being able to use a service track and, if there isn't one, making one. A service track has several functions:
- It makes it easier to get around, which improves the effectiveness of surveillance work.
- It facilitates slope maintenance and allows the use of mechanical means.
- In the event of breaching during a flood, it allows materials (riprap) to be brought in to plug the breach quickly and prevent it getting any bigger.[1]

The track must obviously be built to withstand a certain amount of traffic, including lorries driving about on a partially waterlogged dike body.

The ideal place for a service track is on the dike crest. However, if the crest is too narrow, it can be built on a berm, or even at the toe of the land-side slope. Service tracks on the river side are not practical for in-flood surveillance purposes or for bringing in materials for emergency repairs since they become dangerous, and even impassable, if the flood is a major one.

Service tracks should be regularly maintained to ensure they remain practicable. This basically involves filling in ruts and potholes and ensuring they have a camber to allow rainwater to run off.

5.1.3 Kilometre markers

To help locate features to be noted during inspections and sites where maintenance and repair work needs to be done, it is essential to be able to refer to distance markers positioned on the side of the dike crest. These should be placed every kilometre (KM) or, better still, every hundred metres. In most cases, markers were incorporated at the time of dike or crest roadway construction. If not, they need to be added. They should be clearly visible so that they are not damaged when work is in progress on the dike.

Marker maintenance entails making sure they are still in place and replacing or repositioning them if they have been damaged or knocked over.

---

1. In France, during the Camargue floods of 1994, the majority of breaches that could be reached by lorry were filled with riprap during the flood. Admittedly, they were breaches caused by piping and not overtopping and the water level was still a good distance below the dike crests.

5.2 Vegetation control

5.2.1 Objectives
There are three objectives involved in the control of vegetation:
- Maintaining perfect visibility of dike slopes and bases (to facilitate visual inspections and guarantee their quality).
- Preventing the spread of roots (trees and bushes) through the body of the dike, which not only increases the risk of piping (linear cavities left by rotten roots) but also deforms and breaks up (by mechanical action) any masonry that may be on the surface, such as stone facings.
- Discouraging burrowing animals from making their homes in the dike, by disturbing them (generally shy animals) with the regular passage of service vehicles or machinery and by eliminating covered areas where they can seek shelter.

5.2.2 General objectives
Two objectives apply to the crest and slopes of dikes and a 5 to 10-metre band either side of the base of slopes:
- Keeping grass cover as short as possible.
- Removal of all woody vegetation.

5.2.3 What should be done with existing trees?
When a dike is wooded or has tall, isolated trees, we recommend removing them since: when they eventually die, their root systems will decay, leaving linear cavities where internal erosion (piping) may be initiated when water levels are high. It should also be borne in mind that the roots will continue to decay once the trees have been felled.

At the same time, this implies consolidating the seal of the dike[2], which can be done in one of the following ways:
- Removing isolated trees on the river side, extracting stumps and then clearing out, backfilling and carefully compacting holes.
- Constructing an impervious shoulder on the whole of the river-side slope immediately after felling (followed by stump removal and evening out of the slope).
- Incorporating an impervious cut-off system into the dike (sheet piles or grout diaphragm wall) at the most ten years after felling (the time for roots to decay).

The more or less wooded area between the bank of the main channel and the river-side toe of the dike can be left. It effectively contributes to reducing the speed of the current along the dike and therefore limits the risk of external erosion of the embankment. This vegetation should be kept under control however, notably by felling trees that threaten to fall in the water, taking part of the bank with them and possibly creating an obstruction.

---

2. If it is not possible to consolidate the seal of the dike within a reasonable delay, it's then preferable to postpone felling trees.

5.2.4 Maintenance of grassed slopes

The presence of hardy, well-maintained grass improves a slope's ability to withstand overtopping. The main aim is therefore to keep ground cover homogenous. If need be, more seed can be sown where grass is thin or in poor condition.

Grass needs to be cut regularly to promote healthy growth and maintain good visibility along an embankment. Ideally, it should be cut once a year.

In temperate climates, cutting is best done in autumn or at the beginning of winter for the following reasons:
– The growing cycle has finished by then, cutting will not encourage excessive growth and the embankment will remain clear throughout the winter.
– Birds are no longer nesting.

The cost depends on working conditions (which depend on access): 0.08 euro per square metre if done with a mechanical verge cutter; 0.22 euro per square metre with a portable strimmer.

There is therefore a clear advantage in having service tracks both on the crest and at the toe of the dike so that verge cutters can cover the entire slope.

In addition to cutting, the use of herbicides:
– Means that growth can be limited, making the need for cutting less frequent: mefluidide (cost = 0.08 euro per sq.m), with products such as Green Limit (120 g/l), Embark 120 (120 g/l) and Embark SS (240 g/l) sold by CFPPI.
– Makes it possible to kill all weeds on masonry or drainage embankments: glyphosate (cost = 0.08 euro per sq.m), with products such as Roundup Biovert Aqua, Roundup 360 and Hockey GS2, sold by MONSANTO.

Many products are now available that, if used properly, have very little, if any, impact on the aquatic environment. Public organisations such as, in France, the local department of agriculture and forestry (DDAF) or specialist associations or private bodies can provide good advice on the choice of products.

The introduction of grazing animals is also recommended because the imprints left by their hooves and the natural manure they provide encourage regeneration of herbaceous ground cover. Sheep, rather than cattle are preferable, since heavier animals are more likely to leave deep ruts and tracks.

Even so, care should be taken not to have too many animals and the grazing period should be chosen carefully (damage to embankments from trodden paths, trampling of the ground in very wet weather). We recommend that an official grazing agreement be drafted between the dike operator and the animal farmer. Finally, grazing will probably not mean that mechanical maintenance can be dispensed with entirely, if only because of the plants the animals leave uneaten.

5.2.5 Removal of woody vegetation

Thorough, regular cutting prevents the growth of trees and bushes.

Existing trees and shrubs can be removed conventionally, by cutting down and devitalising roots, in the autumn when the sap is falling (cost = 0.3 to 1.5 euro per sq.m). Otherwise, refer to 5.2.3 above.

Now herbicides are available to kill standing shrubs in a single spraying:
- Either before the leaves fall: fosamine ammonium (cost = 0.04 euro per sq.m), which can be found in Krenite and Krenite Forêt, sold by AgrEvo and AROLE.
- Or during the growing season – triclopyr (cost = 0.02 euro per sq.m) which can be found in Timber and Timbrel (sold by AgrEvo and AROLE) and in Garlon Inov (sold by DAO Agrosciences). Garlon Inov is formulated as an "amine salt" compound, which is more expensive but safer for aquatic fauna than "ester" products (Timber, Timbrel and earlier Garlon 2, 3 and 4 E).

The disadvantage of these methods is that stems, and potentially therefore, ground cover, remain in place until they have biodegraded, unless cut back a few weeks after treatment (this practice also prevents woody growth from sprouting again).

5.3 Burrowing animals

5.3.1 Damage attributed to burrowing animals

The risks and damage caused by the activities of burrowing animals in and around dikes are numerous:
- Development of internal erosion, which may lead to piping (shortening of seepage paths).
- Direct seepage (through-dike burrows or warrens).
- Collapsed areas/unevenness along the crest.
- Mechanical weakening (river banks, river-side slope).
- Destabilisation of masonry, hard facings and roadways.

5.3.2 The main culprits in France and their status as regards French legislation

For all dikes, no matter what distance from the riverbed:
- *Badger*: Timid animal whose sets are roughly 40 cm in diameter. Digs a network of 5-10 tunnels each 8-10 m in length, complete with air shafts.
- *Wild rabbit*: Warrens 10-20 cm in diameter, likes sandy-silt soils, easily identified by its droppings.
- *Fox*: More limited burrowing activity (often lives in sets with or abandoned by badgers).

For dikes and river banks close to the main channel, two large, non-indigenous rodents have adapted to living in France:
- *Coypu*: Native to Central America, very active burrower in climates resembling that of France (fears the cold), makes dens 25 to 40-60 cm in diameter at the entrance and several metres in length. In areas where their population is dense, it is possible to find one den every 0.3-1.5 cubic metres for every 50-60 m of riverbank.

– *Muskrat*: Native of North America, digs a network of tunnels (slightly smaller than the coypu's), always with underwater openings.

Coypus and muskrats sometimes occupy the same territory. Cases of coypus and badgers doing so on "dry" dikes have recently been reported in the Camargue area (Rhône delta). Furthermore, beavers are not listed as animals that cause significant damage to dikes.

The status of burrowing animals in terms of French legislation for the protection of nature:

Wildlife for hunting or not?
Wildlife species that *can be hunted* (in mainland France and Corsica) are catalogued in a decree dated 26 June 1987.

Pests or not?
– National list of species *liable* to be classed as pests (Article R227-5 of the French Code of Rural Law, decree dated 30 September 1988). This list only mentions wildlife that can be hunted.
– Lists of pest species by French "département"[3] (Article R227-6 of the French Code of Rural Law). Fixed by local government by-law passed before 1 December each year (effective from the following 1 January), this list may vary from one part of the "département" to another.

There are specific periods for and means of destroying pest species (e.g. digging out, trapping, shooting) and strict rules for doing so (Article R227-8 and thereafter of the French Code of Rural Law and miscellaneous decrees passed to enforce them).

The following table indicates the status of five species that are of particular interest to us:

Type of animal	Hunting authorised	Liable to be classed as a pest
BADGER	YES	NO
RABBIT	YES	YES
FOX	YES	YES
COYPU	YES	YES
MUSKRAT	YES	YES

The badger therefore differs from the others in that it cannot be classed as a pest.

---

3. French territory is divided into almost 100 "départements", in which the "Préfet" (local governor) represents central government.

5.3.3 Deterrents

Regular mowing, cutting and clearing disturbs wildlife and also prevents the development of dense plant cover, thereby reducing the temptation for certain burrowing animals to set up home (e.g. badger).

The introduction of slope protection has been tested in some areas of France:
- Wire netting (galvanised gabion screen – 60x80, 80x100 or 100x120 double-twist hexagonal mesh) covered with topsoil. Cost = 3 to 4.5 euro per sq.m for the Petit Rhône dikes in the Camargue.
- A facing made of heavy or strong materials. Casings made of interlocking riprap seem to be very effective for this, although their (fairly costly) construction is usually dictated by other constraints (protection against erosion by the current).

Finally, in (parts of) dikes made of non-cohesive materials (gravel, coarse sand), animals are unable to dig tunnels.

5.3.4 Eradication – control of animal populations

In France, whatever the means envisaged, such operations should be negotiated and prepared in association with the relevant services of the department of agriculture and forestry's area offices (DDAF).

a) TRAPPING

French legislation on trapping is strict and is largely based on a ministerial order dated 23 May 1984.

There are six categories of traps. Apart from cage traps (category 1), traps must be approved and marked (except for category 5 – simple deadfall traps) and put in place by authorised trappers[4]. Body-gripping traps *(category 2) have been banned in France since 1995.*

Trappers are also obliged to comply with the following:
- Mandatory declaration made at the town hall.
- Obvious marking of deadly traps (categories 2 and 5).
- Obligatory daily (morning) inspection of traps.
- Record of trapped animals.

The use of cage traps (category 1) is recommended for the following reasons:
- Few statutory restrictions (straightforward declaration at the town hall).
- High selectivity (non-targeted animals are released).
- Highly efficient for intermittent control (e.g. coypu).

Certain animals of a suspicious nature, such as badgers[5], eventually outwit cage traps. In that case, humane stop snares (category 3) can be used, which, like cage

---

4. Authorisation is given by the Prefect after the applicant has attended a training session on trapping.
5. The trapping of live badgers is possible in accordance with articles 9 and 11 of a decree dated 1 August 1986, which allows owners (in possession of an individual authorisation issued by the Prefect, which specifies the ways and means) to catch certain species of wildlife, to keep them temporarily and subsequently release them for the purposes of repopulation.

traps, do not kill the animal and are therefore selective. They can only be used by approved trappers, however.

b) ERADICATION BY SHOOTING

In France, eradication by means of firearms is governed by both national and area regulations.

It requires a "permit to hunt" (which allows ownership and use of hunting weapons) (French Code of Rural Law, article R. 227-16).

In addition to the provisions of the French Code of Rural Law, general conditions for eradication are set by ministerial decrees, which notably specify prohibited weapons, ammunition and ancillary equipment (French Code of Rural Law, articles R. 227-6, 18 and 21; decree dated 1 August 1986).

Eradication periods (R. 227-16, 17, 19 and 20), formalities (R. 227-17, 18, 20 and 22) and places (R.227-17) are specified by regional government by-laws, which follow the provisions of the French Code of Rural Law. Such decrees are valid for a calendar year.

c) HUNTING

Shooting (with dogs and hounds to flush out, stalk and retrieve the quarry) is practised during the open season (Article R224-3 and thereafter of the French Code of Rural Law), the dates of which are set by the Prefect. Hunters must be in possession of a permit and shooting equipment is subject to government regulations on the possession and use of firearms (notably a decree dated 1 August 1986 and order-in-council dated 6 May 1995).

The open season for hunting in mounted or foot packs with horns, hounds and no firearms runs from 15 September to 31 March. It concerns fox hunting (article R. 224-1).

The terrier work season closes on 15 January, although the prefect may authorise the flushing out of badgers for a further period from 15 May (Article R. 224-2).

Underground terrier work is governed by decree dated 18 March 1982. For example, in the Maine-et-Loire French "département", dogs are used for hunting badgers and ferrets for hunting rabbits. Terrier work is also possible for hunting coypus.

d) CHEMICAL CONTROL

For information: In France, it is not permitted to use poisonous substances to eradicate burrowing animals.

e) TO RESUME

Deterrence is always preferable to *the eradication or capture* of burrowing animals. The latter methods – the effects of which are temporary – should be limited to serious or urgent situations, after having previously obtained the opinion and advice of the relevant section of the DDAF, local section of the French national hunting commission (ONC), the area hunting & shooting federation, "lieutenant de louveterie" (technical advisor to the authorities on matters of wildlife control) or the area branch of the approved trappers' association.

Before embarking on a campaign to eradicate burrowing animals, the species involved should be identified, its population density established and the extent of damage caused to dikes and/or river banks evaluated.

If it is necessary to control the population of a (or several) species of animal, insofar as no single method alone is satisfactory, control strategy should ideally be:
– *Selective and non-destructive* to ensure the right species is targeted and, as much as possible, protect the life of the animals; if traps are used, they should not kill the animal.
– *Integrated*, using complementary methods and making efforts to minimise the impact on the environment (prioritise and combine the least traumatising methods).
– *Concerted,* by encouraging collective control, which will reduce the likelihood of re-colonisation spreading from untreated areas, but also by initiating dialogue with local animal protection organisations (to study, for example, the possibility of releasing captured animals into areas where they will not be detrimental).
– *Programmed* in time and place (management plan).

In conclusion, the following table takes each of the five species implicated in dike damage and summarises the methods that are authorised in France[6] and that we recommend for controlling populations:

	Trapping	Shooting	Hunting, incl terrier work	Recommended methods
BADGER	YES (individual authorisation from the prefect)	NO	YES	1 Deterrence: slope maintenance (timid animal), mesh covering 2 Trapping (Humane stop snare)
RABBIT	YES	YES	YES	1 Deterrence: mesh covering on slopes 2 Shooting, terrier work or trapping
FOX	YES	YES	YES	1 Deterrence: mesh covering on slopes 2 Trapping (cage trap)
COYPU	YES	YES	YES	1 Shooting or trapping 2 Terrier work
MUSKRAT	YES	YES	YES	1 Trapping

In point of fact, although different countries have their own particular species of burrowing animals (in Vietnam, for example, it's termite nests that cause damage to

---

6. It should be remembered that, in France, the use of poisonous substances is forbidden.

dikes!) and their own legislation, it should be relatively easy to adapt the principles outlined above in order to control such populations.

5.3.5 Curative measures

a) Injection of burrows

As a curative measure, the injection of hardening liquids may be considered for filling in tunnels, which are a source of dike seepage and weakening.

To our knowledge, this technique was used in France on a CNR Rhône dike (*cf*. fig. 6) at the Péage-de-Roussillon site in 1996, after it had been weakened by the presence of rabbit warrens.

The composition of the cement-bentonite filler was as follows for 1 cubic metre (density of 1.56):
– 125 kg of cement CPJ 32.5 R.
– 735 kg of sand 0.1-0.3 mm.
– 44 kg of bentonite.
– 660 litres of water.

Before injection, the rabbits were caught (although not all of them, despite the efforts of the local hunting club). The mixture was injected from a pump truck (in which it was also mixed), using an 80 mm diameter hose.

Altogether, about 2 km of dike were treated (110 warrens) with 16 cubic metres of mixture, working out at 150 litres per warren. The work required six times more mixture than expected...

The cost of the operation amounted to approximately 6,100 euro per km. It did not prevent damage from appearing again next to the injected warrens (excavation of new warrens), which points to the obvious need to combine curative measures with deterrence.

In the Camargue (Rhône delta), burrow entrances are also filled using basic earthworking equipment (mini excavators, sacks of earth, clay fill and compaction hammers). It is essential to catch the animal first since, if caught prisoner in the burrow, it will inevitably dig another way out. The effectiveness of such efforts – i.e. in providing a seal against a hydraulic head – is not known.

On the other hand, and as far as we know, the results of trials using injected expanding foam were inconclusive since the foam shrinks as it hardens, meaning that an impervious seal cannot be guaranteed with this type of repair.

A more simple method is to use mechanical equipment to excavate the area of the dike that has been affected by burrowing and re-constitute the section by filling it with the re-compacted extracted material, as long as it has the appropriate properties and water content for this purpose.

b) Sealing techniques

To restore the seal inside dikes that have been excavated by burrowing animals, the following techniques can be envisaged:

– A diaphragm wall or sheet pile cut-off running along the dike.
– An impervious shoulder on the river side.

The advantage of diaphragm walls and sheet pile cut-offs is that they definitively resolve the problem of a dike's internal impermeability, even if burrowing continues; wildlife will obviously not be able to tunnel through sheet piles or through a diaphragm wall once the mixture has set. On the other hand, these techniques do not help consolidate a dike's river-side slope, which may be problematic if burrow development (which is bound to continue) especially concerns and weakens that part of the dike (e.g. burrows of aquatic rodents such as the coypu). Such techniques are expensive:
– 90 to 140 euro per sq.m for sheet pile cut-offs.
– 45 to 105 euro per sq.m for a grout diaphragm wall made with a mechanical shovel.
– 30 to 55 euro per sq.m for a thin grout diaphragm wall.

The implementation of appropriate deterrent measures is often needed in addition to curative techniques to protect re-constituted embankments and untreated sections from further "attacks".

5.4 Slope protection and walls

5.4.1 Maintenance of masonry facings

There are three main causes of damage to protective masonry facings on the river-side slope:
– Deterioration of ashlars that were of an inferior quality at the outset.
– Deterioration of the mortar joints that form the bond between ashlars.
– Dislodgement of the toe of the facing.

Since the first cause only concerns a limited number of ashlars, repairs are easily carried out by replacement of deteriorated ashlars. New ashlars should be of dense, hard stone that is not affected by immersion in water. They should be shaped to fit as exactly as possible. If need be, fragments of stone can be hammered in to fix them in place.

If damage to ashlars affects large parts, or the whole, of a slope, it is because their source has been poorly chosen and major repair work must therefore be envisaged.

With time, weathering of mortar joints is inevitable, especially on old structures. What usually happens is that the mortar is broken down by physico-chemical mechanisms, plant life is then able to take a hold in the spaces between stones and, if nothing is done about it, ashlars are eventually dislodged by roots (trees growing up through stone facing is unfortunately not a rare sight).

Besides controlling vegetation as described above, periodic re-pointing also needs to be programmed (every 30 to 50 years if the facing was well done in the first place and properly maintained since).

The operation consists of removing all old joints to a depth of 5 to 6 cm, cleaning them out with compressed air or pressure-sprayed water and re-doing the joint, which is smoothed into a hollow shape or concavity when viewed against the adjoining ashlars.

Since stone facing is not generally expected to act as an impervious barrier, to ensure stability when flood waters retreat, joints should not be continuous so that interstices are left for the eventual dissipation of uplift.

Dislodgement at the toe of facings occurs frequently because the bottoms of many watercourses in France have a tendency to gradually deepen.

The toe of the facing, which was originally buried or protected by permanently submerged piles, is then exposed in an area of rising and falling water levels and localised dislodging quickly sets in.

The remedy lies in putting in a new system of linear protection (frequently with sheet piles) and rebuilding the abutment on the cut-off structure.

5.4.2 Protection of riprap embankments

River-side slopes are sometimes protected from erosion by a layer of riprap.

The transition layer between the embankment earthfill and the riprap is usually a geotextile and less frequently a granular transition layer, although there is often no transition layer at all.

Damage, its causes and conceivable repairs are as follows (not an exhaustive list):

Damage	Probable causes	Conceivable repairs
Torn geotextile	– Geotextile not strong enough – Roughness of underlying layer – Damage to geotextile when riprap put in	– Remove the riprap and geotextile in the damaged area, smooth out rough points of underlying layer, put in a new and stronger geotextile (non-piercing geotextile) with a minimum 0.5 m overlap on top of the previous geotextile (which is left in place), and put riprap back in.
	– Wildlife holes	Same as above and add wire netting
Physico-chemical degradation of geotextile	Geotextile directly exposed to sunlight	Same as above. Make sure the riprap layer is packed closely enough to prevent solar radiation reaching the geotextile
Riprap degradation	Crumbling stone, unsuitable geological properties	Partial or total repair of the riprap layer depending on the extent of the problem

Damage	Probable causes	Conceivable repairs
Riprap slide	Foundation anchor block too small	Increase the size of the anchor block. Add to the layer of riprap at the head of the slope to be protected
	Too much gradient	Increase the layer of riprap (by also widening the anchor block) to reduce gradient. Increase riprap stability by bonding with masonry (faced riprap)
Riprap swept away	Riprap not large enough, layer not deep enough	In-depth diagnosis and rescaling of protection

5.4.3 Maintenance of masonry walls

It is quite common to see plant life taking a hold on the faces of masonry structures. Crevices in the jointing provide a perfect place for seeds to collect and find the moisture they need to flourish.

Root development can cause considerable damage to joints and mortar and, in extreme cases, can even lead to the upheaval of facing stones.

Therefore, masonry work also needs to be kept completely clear of vegetation by uprooting plants as soon as they appear. As in the case of stone facing, pointing deteriorates and needs redoing from time to time (see previous section). Annual uprooting is recommended.

Other types of damage to mass masonry are of a mechanical origin (differential settlements, global instability) and appear as cracks affecting the entire structure. Although some repairs can be carried out by the operator, the extent of the damage needs to be assessed, a diagnosis of the causes made and an informed decision on repair techniques taken[7], all of which are matters for a competent technical service.

5.5 Toe protective works

Preliminary comment: In nearly every case, damage to toe protective works is serious. It is thus a matter of repair rather than maintenance and requires a preliminary specialist diagnosis.

---

7. Reference can be made to French standard NFP 95-107, entitled: "Repair and upgrading of masonry work – specifications for techniques and materials".

5.5.1 Timber piles

More often than not, where linear protective works exist, they consist of timber piles. As long as these piles remain under water, they are extremely durable. However, because of a relatively common phenomenon in France whereby the main channels of river beds tend to gradually deepen, piles may be exposed to the air when water-levels are low. Their useful life is then considerably reduced since the wood decays, breaks up or disintegrates.

Correcting this is not simply a matter of maintenance and very often involves the replacement of timber piles by metal ones. Timber piles can also be used provided their heads are lowered to ensure they always remain submerged, and that the appropriate type of timber is used (chestnut, oak, azobé, etc.).

Whatever the case, a preliminary study needs to be conducted to include a diagnosis of the damage and its causes, the choice of repair techniques and a detailed work programme.

5.5.2 Sheet piles

Sheet piles are recent installations that age principally through rusting. Areas of significant rusting should be tested for residual thickness (non-destructive ultra-sound measurements rather than drilling). Repairs involve considerable work and should be preceded by an in-depth study.

Badly distorted sheet piling is another case for an in-depth study by a specialist in geotechnics.

5.5.3 Gabions

Gabion anchor systems are less frequent.

There are two reasons for gabion wire netting to break:
– A localised break is often caused by a sudden impact or bending that has led to chipping of the galvanised wire. If attended to rapidly, repair is simple, a new section of galvanised wire being introduced as shown in Fig. 14.
– Extensive corrosion may appear in areas where the water level fluctuates, producing whole lines of broken wire, with the result that the contents of the cage spill out. Repairs should be carried out as soon as generalised corrosion is noticed and, in any event, before multiple breaks set in. The corroded netting should be covered with a new section of galvanised netting, which needs to be firmly fixed all the way around. If necessary, a mortar facing can be added to areas especially exposed to corrosion.

5.5.4 Riprap

Toe protective works can also take the form of a riprap abutment (anchor block or staunching wall). If the abutment suffers damage (riprap swept away, large areas driven in), a preliminary diagnosis needs to be carried out (extent, causes, solutions) before any repair work should begin.

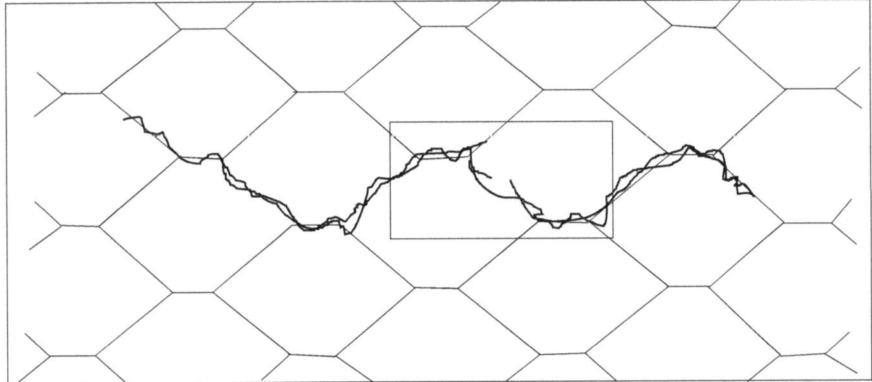

Figure 14. Repair of broken wire on a gabion cage

6 — THE BASIC PRINCIPLES OF DIKE DIAGNOSIS

Introduction: from rapid diagnosis to in-depth diagnosis

The purpose of any diagnosis of a civil engineering structure is to evaluate how safe it is, identify its weaknesses, faults and malfunctions and describe the upgrading work to be done to remedy those anomalies.

Dikes are a special case in that they are long, linear constructions about which detailed information is often lacking (building plans generally not available), which are sometimes badly maintained and subject to infrequent, though intense, hydraulic and mechanical stress – all factors that tend to make diagnosis difficult and, therefore, costly.

We believe it to be appropriate to distinguish two levels (or stages) of dike system diagnosis:
- The first is *rapid diagnosis*, based on the following programme of work: clearance of ground cover (see section 5.2), a topographical survey of the structure and its immediate surroundings (at least to include regularly-spaced longitudinal and transverse profiles

using, if possible, 1:500 maps – see section 6.3) and an initial visual inspection (see Chapter 3).
– The second is *in-depth diagnosis*, which incorporates all stages of investigation outlined in the following parts of this chapter.

The advantage of making this distinction is to encourage – at least for most French dikes – the immediate and minimal implementation of the first stage – *rapid diagnosis* – in order to obtain a low-cost evaluation of the state of the structure and bring together all the conditions required for its subsequent regular surveillance and maintenance. Once a rapid diagnosis has been done, an *in-depth diagnosis* can then be conducted, and could be restricted to the critical sectors identified in the first diagnosis.

6.1 Identifying failure risks and adapting the diagnosis

The various mechanisms responsible for the formation of breaches that were examined in Chapter 2 will serve to determine how to diagnose a segment of dike and the most efficient way to conduct the survey. The main principle is to establish a hierarchy of hazards to which a particular segment of dike appears to be most exposed. For example, for a dike that verges on the main channel, diagnosis will concentrate particularly on the risk of scouring. On the other hand, a study of geotechnical failure probabilities would not usually be a priority for a wide dike topped by a roadway.

a) The most common failure mechanism is *overtopping*.
The risk largely depends on the magnitude of the flood and it is useful to be able to use up-to-date hydrological and hydraulic surveys that give the water levels for different periods of flood recurrence. An accurate topographical survey of the longitudinal profile of a dike is indispensable for evaluating the risk of overtopping. It is then possible to establish a datum event, which is the extreme event against which dikes are expected to protect the flood valley.

b) The second mechanism involves *erosion and scouring* of river-side slopes, and particularly their toes, of dikes bordering a main channel.
For many watercourses in France, this risk has probably increased since the 19th century due to main channel deepening (because of the extraction of materials in the past) and to the ageing of slope protection (facings and their pile foundations). Careful visual examination, above and possibly below the water line, and a morphological and hydraulic analysis are appropriate measures to consider in diagnosing this type of risk.

c) The risk of *internal erosion* (or *piping*) becomes greater the longer the flood lasts and the older the dike is (burrows and warrens, dead tree roots, through-dike structures, differential settlements).
Diagnosis will be largely based on a very detailed visual examination, the locating of all conduits, pipes and tunnels running through the dike and information obtained from local residents (location of leaks during floods). Observations made during flooding will provide further valuable information. Geotechnical exploration aimed at

characterizing the heterogeneity of materials and the permeability of the dike and its near foundations will make it possible to assess the risk of piping and may be based on hydraulic modelling of the levee.

d) Although *global instability* appears to be a marginal factor in terms of the breaches observed, it should not be neglected, particularly for levees with narrow crests and steep slopes, as well as in areas of former breaching and for masonry dikes.
The heterogeneity of many levees makes it difficult to accurately diagnose instability. It seems to be more logical to carry out parametric studies, based on qualitative investigations of materials and on data obtained from any recent investigations of similar sectors, in order to arrive at safety margin coefficients and especially examine the improvements in stability afforded by different upgrading solutions.
A detailed study of archive data should make it possible to locate the majority of old breaches and it is on these areas that specific geotechnical explorations should then concentrate. For recent structures (reinforced concrete walls, for example), the first step is to find the dossiers that correspond to structures that have been built, along with their definitive scaling details (calculations, plans) or, failing that, preliminary designs.

e) Lastly, all special features and appurtenant works (berms, freeboard walls and ridges, spillways, pipes, conduits, etc.) may constitute vulnerable points. Visual inspection will be the main, and often the only, way to identify them and assess their condition. More specific investigations can be carried out subsequently if need be.

6.2 Historical research

The historical background of a dike is an essential part of dike diagnosis and even constitutes the first stage. Information should ideally be obtained on:
- The history of the construction, upgrading and management of the dike.
- The composition of embankments and particular features.
- The location of old breaches, which, far from being random, is largely determined by the geometry of a dike-protected channel and the exposure of levees to water action during flooding.
- Historical floods of watercourses.
- Changes in the occupation of the space in the dike-protected channel and the protected valley (protected assets and interests).

The contribution that can be made by historical analysis is therefore evident.

It is therefore appropriate to:
- Find and be aware of all available documents – district and municipal archives, operator's dossiers and plans, aerial photos, previous studies, files on upgrading work, etc.
- Locate historical breaches – of great importance in our opinion.
- Enumerate and locate the signs of historical floods.
- Compile an inventory and review of river materials extracted from the area.

6.3 Topography

6.3.1 Objective

There are three objectives in conducting topographical investigations on the dike:
– To establish the link with the river surface profile during flooding.
– To identify transverse profiles to be examined during geotechnical studies.
– To provide a dike reporting and surveillance tool.

The following sections detail these objectives and the means to be implemented to attain them.

6.3.2 Longitudinal profile of dikes in relation to highest flood-water levels

We have seen how overtopping is a major cause of breach formation, at least for earthfill dikes. It is possible to evaluate this risk by studying the river surface profile during flooding in relation to the profile of the dike crest.

We recommend drawing a longitudinal profile at maximum intervals of 20 or 25 metres along the top of the dike on the crest platform, and a second profile along the top of the freeboard feature (if it exists), in order to calculate the freeboard available in relation to the flood peak of the design flood and to highlight segments where this freeboard would be inadequate.

Correlating datum water levels and the dike's geometry demands that profiles be accurately set against the same reference systems for altitude (e.g. NGF in France) and KM.

6.3.3 Transverse sections

During flooding, the function of a dike is to maintain the difference in water level between the dike-equipped channel and the protected valley. The hydraulic head may be as much as 5 or 6 metres during exceptional flooding along the large levees of big French rivers (e.g. Rhône, Loire and Garonne) and more often 2 to 4 metres for smaller watercourses.

Failure mechanisms to watch out for are piping (retrogressive internal erosion of the dike or its foundations) and instability of the land-side slope during the flood and of the river-side slope when the water retreats. In either case, risk analysis demands possession of the dike's transverse profiles.

Transverse profiles also provide basic data for planning and identifying any necessary upgrading work.

A transverse section should be surveyed every 100 to 200 metres in homogeneous areas and every 50 to 100 metres in complex areas, including a sufficiently wide band on the river side and land side (about ten metres on either side). Each transverse section should contain at least 8 to 12 points depending on the size and complexity of the structure. Depending on its configuration (notably, the presence of particular features), additional points may be necessary.

6.3.4 Topographical map

Drafting a 1:500 or 1:1000 topographical map is especially useful when the dike has numerous particular features. Such maps are also invaluable for surveillance and maintenance work.

The map then serves as an aid to visual observations, which form the basis of dike surveillance. Recent experience has shown that the cost of such a survey comes to between 1,500 and 3,000 euro per kilometre for a segment of several kilometres.

It is worth linking the map to a reference system (e.g. the Lambert grid in France) with a view to its future integration into a geographical information system (G.I.S.) and saving it in a format that is compatible with hydraulic modelling software (RIVICAD, for example).

6.4 Visual inspection

Because visual inspection is of paramount importance in dike diagnosis and surveillance, an entire chapter (Chapter 3) has been devoted to it.

6.5 Hydrological and hydraulic survey

A hydrological survey involves determining the nature of floods with different recurrence intervals (flow rates, duration and frequency). It is based on watercourse flow rate measurements taken at stream-gauging stations, together with information on historical floods. Significant changes in land use in catchment areas (dense urbanisation, extensive reforestation programmes, etc.) or large-scale upstream developments (flood-control dams) are liable to modify flood-water regimes (especially during medium-intensity floods) and may necessitate the updating of previous hydrological surveys.

Hydraulic surveys are used to convert the results from hydrological studies into flow lines for ten-year, thirty-year and hundred-year (or more) floods. They require a knowledge of the detailed topography of the stream bed (costly) and the implementation of an hydraulic model. In most cases, a steady-state, mono-dimensional model is sufficient.

Historical flood flow lines may provide enough information, dispensing with the need for the hydrological and hydraulic surveys mentioned above, provided that:
- Historical flooding has not led to dike failure.
- Stream bed modifications (longitudinal profile, new embankments, changes in floodplain land use) do not lead to any significant change in flow lines at equivalent flow rates.
- The hydrology of the catchment area has not changed significantly.

Comparisons of the flow lines for different flood recurrence intervals against the longitudinal profile of a dike make it possible to define the **maximum design flood** (datum event) – that is, the most extreme event against which the dike is expected to protect the valley.

The survey should be completed with an analysis of scenarios of exceptional flood peaks and associated phases of retreating water (spillover, filling and draining times of the flood spreading plain, operation of spillways, evacuation works, flap gates, sluices, etc.).

6.6 External erosion and scouring: the geomorphological approach

Research into the causes of dike failure during major historical floods shows that breaches tend to occur in the same place as, or near, old breaches. This uneven distribution can be explained by the fact that, in certain segments, a dike attempts to counter a powerful morphodynamic change in the watercourse, which leads to chronic weakness.

The purpose of a geomorphological (or morphodynamic) approach is to identify areas of historical risk and current unfavourable developments.

In the most common cases of localised channel narrowing, the process involved is hydraulic (overtopping at the point of narrowing) rather than morphodynamic. However, a morphodynamic approach provides information about other causes of failure:
– Failure at points where channel gradient changes.
– More frequent failure when the dike is in immediate contact with the main channel.
– Weak points on the outside of meanders.
– Breaches in sections of watercourses with multiple channels and the presence of islands colonised by plant life.

Generally speaking, diagnosis should make it possible to identify past and future changes in the channel: meander soil creep, lateral displacement of islets, lowering of the river bed, hard spots and breaks in slope profiles.

In this instance, the appropriate spatial scale for analytical purposes would be several kilometres upstream and downstream of the dike, including the diked bed and the parts of the valley on either side of the dike. A knowledge of old flood branches or propagation channels provides information on sectors the most exposed to scouring.

Theoretical analysis of the morphodynamics of the watercourse is based on knowledge of its hydrology, sedimentology and morphometric properties. The analysis is seg-mented on the basis of knowledge of longitudinal slopes, alignment of the water-course, anthropic action (particularly material extractions between or behind dikes). A comparative study of the watercourse's longitudinal and cross-sectional profiles is made using existing documents. Modifications to the course of the river are assessed in terms of a coefficient of meander sinuosity and radius of curvature and in terms of hydrographic network density. Changes in the scale and rate of colonisation of islets by plant life are noted over time.

Resources for carrying out this work are:
- Old maps and low-water and historical flow lines.
- 1:25,000 or 1:50,000 national survey maps at different dates.
- Bathymetry on various dates.
- Aerial photos taken at different periods.

Field analysis provides a means to refine and qualify the findings of theoretical analysis. Watercourse network surveys are carried out on pre-determined segments. The survey provides information on the sedimentological component and the processes of change in the river bed. Types of erosion (in blocks, by crumbling or chipping) and deposits (convexity, broadening – hydraulic load losses, exogenous reasons, obstructions) are defined. Apparently stable or unstable profiles are identified and possible developments analysed: acceleration in meandering (rapid extrados erosion, area of convexity in the process of re-vegetation), lowering of the river bed (overhanging riparian growth, obvious scouring marks on structures, breaks in slope profile, etc.). River bed granulometric change is related to watercourse slope and alignment: it is therefore a good idea to plan to collect samples of bed materials for laboratory analysis in order to specify the river dynamics, highlighting the phenomenon of granulometric sorting in particular. Finally, regular bathymetrical surveillance can be introduced for areas apparently exposed to scouring.

6.7 Geotechnical diagnosis

6.7.1 A few notions about safety in relation to dike composition

In general, dikes are limited in height and their stability is, in principle, less critical than that of bigger constructions such as dams. However, in contrast with dams, dikes are not under hydraulic load in normal circumstances and we cannot rely on observations of their everyday behaviour to make assumptions about their safety in extreme situations. Which leads us quite naturally to recommend applying wider safety margins to this type of structure.

Prior to a more in-depth study, it is possible to give a few indications in order to make an initial assessment of a dike's safety in geotechnical terms.

We can start by looking at dike slopes.

River-side slopes with gradients in excess of 1(vertical): 3(horizontal) present a risk of instability when flood waters retreat due to the waterlogging of river-side fill materials. The same applies on the land side in terms of stability during a flood, as that area becomes progressively saturated.

Crest width and the gradient of slopes will tell us about width at the base of the dike, which dictates average hydraulic gradient inside the embankment. For Loire levees, studies conducted in the 1960s led to the proposal of upgrading solutions on the basis of:

$$L > 8\,H \qquad \text{L: } \textit{width at the base}$$
$$\text{H: } \textit{height of the dike.}$$

This still seems to be a reasonably valid basis on which to make an initial assessment of the geotechnical safety of an earthfill dike.

The nature of the component materials also has a bearing: a dike made from sandy materials is at greater risk than a compacted silt dike because of its permeability and the speed at which a high piezometric head can develop in the body of the structure.

The presence of a draining device (blanket at the land-side toe) or simple zoning of materials (coarser materials on the land side) is an important factor in terms of safety.

6.7.2 Conditions for defining a programme of exploration

An appropriate programme of geotechnical and geophysical exploration can only be worked out – and conducted – after completion of a minimum number of stages in the diagnostic study:
– Historical research (see section 6.2).
– A detailed, large-scale topographical survey, 1:500 or 1:1000 or at least a sufficiently accurate survey of the longitudinal and transverse profiles (see section 6.3).
– A visual inspection if possible (see section 6.4).

The historical research stage, which integrates analysis of existing studies, provides preliminary information on the geological environment, as well as on dike composition and incidents – sources of discontinuity – that have taken place (breaches and/or upgrading work). Topographical surveys are useful for locating geotechnical soundings as well as for drawing up exploration profiles and, combined with visual inspection, contribute to the initial identification of particular dike features or segments.

The rational use of the results from these three stages will make it possible to:
– Choose the most appropriate geotechnical exploration methods for the case in point.
– Help pinpoint locations for such explorations.

It is worth noting that geotechnical work quickly becomes an expensive budget item, since the unit price of certain boreholes or tests is high (e.g. core drilling), and that geophysical exploration devices, which may initially appear to be a cheaper option, may produce results that are of little interest or even unusable if the device employed is unsuitable for the job in hand; which justifies the attention that should be paid to preparing exploratory programmes.

Solutions envisaged for *upgrading work* may also govern choices relating to the content of geotechnical explorations. For example:
– Consolidation using a downstream (land side) draining shoulder requires studying the grain size distribution of the materials that make up the land-side slope of the dike and the foundations of the shoulder in order to satisfy the filter rules (see Appendix 2) that ensure that internal fine materials are not washed out towards the drain.
– Consolidation using an upstream (river side) impervious shoulder likewise requires studying the grain size distribution of river-side slope materials in order to satisfy the filter rules, but also requires knowledge of the shoulder's mechanical properties to be able to verify stability of the structure when flood waters retreat.
– Consolidation using a diaphragm wall requires reasonable knowledge of the foundation materials (especially their permeability) in order to set the anchor block at the right level.

6.7.3 Elements for working out a programme of exploration

It is not within the scope of this handbook to go into the details of the geotechnical diagnosis of dikes, which is the work of specialist design offices and needs to be tailored to each particular case (nature and configuration of structures). For further information, *"Dike diagnosis methodology as applied to the levees of the middle Loire"* guidelines, Cemagref Éditions, March 2000, may be of use.

However, as a general guideline, we can mention a few aspects to be included in a programme of geotechnical exploration for a segment of fill dike:

a) CONTINUOUS GEOPHYSICAL EXPLORATION
In principle, continuous exploration has two objectives:
– To provide a spatial vision of the composition of a dike and assess its degree of heterogeneity – on the condition that the instrument(s) used is(are) calibrated beforehand and that a cross-analysis is made with the results of systematic soundings (see B).
– To highlight particular points or segments that are liable to be missed during occasional, systematic borehole work.

Choosing the geophysical tools to employ is not necessarily obvious. To help in choosing a geophysical method, we recommend applying the following two principles:
– Simultaneously applying two longitudinal methods that are based on complementary principles and/or do not measure the same parameters.
– Prioritising methods that, in a single operation (longitudinal profile), can explore deep enough to reach the dike's foundations.

In view of the volume and quality of feedback anticipated, the cost of continuous geophysical exploration should not exceed 2,300 euro per kilometre.

For further information about geophysical exploration methods applied to dikes, please refer to the guideline "Geophysical and geotechnical methods for diagnosing flood protection dikes", available in English and published by "Quae Editions".

b) Intermittent geotechnical surveys

These surveys, which are spatially intermittent in that they are repeated at appropriate intervals along the length of the dike in question, seek to accurately, if only in certain places, characterise the composition, and one (or several) important property (or properties), of the dike. The information obtained will also contribute to the indispensable refinement of previously-implemented continuous geophysical methods.

The results of the previous geophysical exploration help to determine homogeneous segments where geotechnical surveys are to be conducted.

Intermittent geotechnical surveys basically consist of geotechnical sounding (penetrometer, core or destructive drilling with piezometer, Perméafor) and geotechnical testing (Lefranc permeability tests, phicometer and soil identification).

The global cost of intermittent geotechnical surveys should not exceed 3,800 to 4,600 euro per kilometre for soundings taken every 200 metres or so.

c) In-depth surveys of particular features or segments

These surveys are conducted at points or segments of dike where continuous or intermittent geotechnical surveys or visual inspections have highlighted anomalies or particular features liable to reveal the existence of a weak point in the dike. Because of their nature, it is impossible to draw up a model programme for such surveys, which must be tailored to individual circumstances.

Payment for such specific services can only be made on the basis of a unit-priced bill of quantities, which requires that prices cover a broad range of services.

6.7.4 Cost

Excluding special surveys of unusual segments, the cost of the geophysical and geotechnical prospecting method recommended above amounts to approximately 6,100 to 7,600 euro per kilometre.

This item therefore represents a large proportion of the global budget, which underlines the need to carefully prepare the exploration programme.

6.8 Numerical modelling

Numerical modelling is now widely used in geotechnics. Improvements in the computational ability of modern computers and the development of more user-friendly specialist software programmes mean that a whole range of structural loading hypotheses can be tested rapidly on a given structure.

Though useful, such tools nonetheless have two major limitations:
– Any model is an intellectual simplification of the real situation, which is based on the more or less complete representation of a few physical phenomena and their interactions (including boundary conditions).

– The quality of modelling results depends directly on the quality and representative character of the data used to set the model's parameters.

On the first point, we can consider that, being relatively simple structures, dike analysis does not require highly sophisticated models and that many tools used widely in engineering could be considered suitable for use. On the second point however, modelling proves to be limited in that dikes are heterogeneous and certain model parameters are difficult to obtain in a representative and reliable fashion (mechanical properties in particular).

In our opinion then, dike modelling should be carried out by:
– Referring whenever possible to the results of previous studies before embarking on any new calculations.
– Prioritising simple models, the parameters and boundary conditions of which can be relatively easily fixed.
– For dike diagnosis, systematically checking the sensitivity of results by varying the data within ranges determined by the results of exploratory surveys or by other studies.
– Using models to compare a variety of upgrading solutions and/or to optimise their design.

The purpose of *internal hydraulic modelling* carried out in a steady state with a parametric study of permeability values is to obtain the internal piezometric head to be taken into consideration in mechanical modelling in addition to the hydraulic gradients used to evaluate the risk of piping (see section 6.1.c).

Geomechanical modelling is carried out using simple two-dimensional models based on circular or plane failure mechanisms, as part of studies into the overall stability of the dike (see section 6.1.d). It is best to opt for a parametric approach, given that one of the major advantages of mechanical modelling is to assess the improvements afforded by upgrading and to compare different solutions.

6.9 Evaluation of vulnerability

When carrying out diagnostic research, it is generally good practice to include an assessment of the infrastructures and human activities that would be affected in the event of dike failure or malfunction.

A brief assessment of the consequences of dike failure should be made so as to classify segments being studied in order of priority and to gear diagnostic and upgrading methods to the vulnerability of the protected area as necessary.

Vulnerability is evaluated[1] according to the following criteria:
– Land use (urban, periurban, industrial, agricultural, etc.).
– Size of the protected population.

---

1. Vulnerability should be assessed in cooperation with government authorities, especially those responsible for formulating risk prevention plans and land use plans.

– Communication channels and infrastructure under threat (roads, railways, channels, buried pipework, etc.).

and is graded according to vulnerability:
– (1) – Low to medium vulnerability.
– (2) – High vulnerability.
– (3) – Very high vulnerability.

6.10 Prioritisation of risks

Risk results from a combination of hazard probability (unforeseeable turns of events) and vulnerability (importance of human interests liable to suffer the prejudicial consequences of such events). This risk is evaluated for a given flood level, which is usually associated with a datum recurrence interval or historic event.

Failure probability is evaluated on the basis of conclusions drawn from diagnosis, which seeks to classify each segment of dike according to a category of failure probability:
– (1) – reliable dike in terms of the reference event (flooding).
– (2) – dike with a low degree of failure probability.
– (3) – dike with a high degree of failure probability.

The global failure probability of a particular segment is the failure probability corresponding to the failure mechanism or degradation (overtopping, scouring, internal erosion, etc.) most likely to occur.

Evaluation of the *risk* associated with a particular segment is a combination of that section's *failure probability* and the *vulnerability* of the protected area. It is possible to give a score that could be, for example, the mathematical product of the probability and vulnerability scores.

A suitably-scaled (1:10,000) cartographic approach is recommended for conclusions about risk analysis. It should show:
– Division into homogeneous segments.
– Grading by segment of the probability of malfunction and failure.
– Vulnerability by zone of protected areas.
– The category of risk associated with each segment.

6.11 Solutions for upgrading dikes

A study of upgrading solutions constitutes the last stage in the diagnosis of a civil engineering structure, the aim being to correct any observed or suspected damage in a suitable manner. It may be useful at this stage to give consideration to the phasing of work to be carried out, taking into account the results of risk analysis as presented in section 6.9 and 6.10.

6.11.1 Principal options for upgrading depending on technical situation

A rapid description of the main options for dike upgrading in response to various types of threat to dike safety permits the identification of a certain number of constraints to be reflected in the specification of services that the design office tasked with the diagnosis should integrate into its engineering mission. For example, as already indicated in section 6.7.2, conceivable solutions for upgrading may determine the content of a part of geotechnical surveys to be included in the diagnosis mission.

Using the (most common) example of *fill dikes*, we arrive at the following table on the basis of the possible failure and damage mechanisms described in section 6.1:

	Upgrading options	Notes: aims of upgrading, field of application/constraints
Overtopping	Removal of low points or raising of the dike	Adaptation of crest level paying attention to the seal of the freeboard feature
	Upstream spillway	Limiting of flood peak height in the front of the dike
Erosion or scouring of slopes	Protection of the toe of the dike on the watercourse side	For areas in direct contact with the watercourse (risk of initiating movement of the alluvial base)
	Protection/facing of the river-side slope	In addition, or not, to protection at the toe
Piping	Drainage shoulder on the land-side slope	When there is sufficient space – the need to respect filter rules at the surface of contact between the drainage shoulder and the dike slope
	Impervious facing on the river side or diaphragm wall in the body of the dike	Limiting of seepage runoff and/or increase in the length of seepage paths
	Control of burrowing animals	Capture of animals, protective netting, etc.
	Treatment of particular features	Conduit and pipe crossings, buildings, cellars, ... when they are the source of a risk of piping
Global instability	Shoulder on land side and/or river-side slope	Depending on the side of the dike concerned (generally due to too much slope gradient)
	Impervious facing on river side or diaphragm wall in body of dike	To stabilise the land-side slope by pushing back the line of saturation in the dike

Looking at this table, we can see that certain types of upgrading can effectively prevent several distinct failure mechanisms, such as land-side drainage shoulders or impervious diaphragm walls in the dike body.

Besides this, the feasibility of some upgrading work is subject to conditions. For example, the creation of a shoulder on the land side is only possible if there is sufficient space at the toe of the dike (which will not be the case if buildings are embedded into or in close proximity to the slope in question).

6.11.2 Objective to be set at the design stage of upgrading work

The objective to be indicated when studying "upgrading solutions" is to arrive at a design – at least at the preliminary design stage – for all the works needed to improve dike safety in a flood situation, the occurrence of which is chosen as a function of the importance of protected human interests and the cost of protection work. By requiring an upgrading plan worked out to the preliminary design stage, it is in principle possible to ensure that diagnostic studies will produce operational conclusions that can be immediately utilised by the owner or operator's service with a view to scheduling work (if not, it is almost inevitable that a second study will be needed, with the additional cost entailed).

By formal definition of the engineering mission with the drafting of a preliminary design, this type of objective calls for:
– The comprehensive identification and evaluation of all mechanisms responsible for damage or failure likely to impact the dike in its current configuration. Hence the need for a relevant in-depth diagnosis (refer to previous stages in the procedure).
– An inventory and comparison of all theoretically possible upgrading solutions for offsetting the dike's potential deficiencies, in the form of a preliminary scaling exercise.
– The selection, justification according to technico-economic criteria and calculation of the recommended solution(s), after having integrated all practical constraints, whether from the point of view of the director of works (e.g. space restrictions, phasing dictated by budget restrictions, etc.) or of the technical elements previously identified or in the process of being studied.

Subject to all study initiatives recommended in this handbook being applied, the objective of arriving, by in-depth dike diagnosis, at the formulation of a relevant preliminary design for upgrading work appears to us to be completely realistic, at least in the majority of cases.

6.11.3 Justifications and technical constraints to be taken into account

A number of justifications and technical constraints are worth taking into account at the design stage of structures. Below are those we feel to be the most important and which we recommend systematically to be mentioned in the specifications of contracts with design offices.

a) DESIGN OF STRUCTURES

The use of simple internal hydraulics or geomechanical models helps in assessing the safety of a dike in its current state (see section 6.8). The same tools can be employed at virtually no additional cost to evaluate the condition when consolidated. Despite the imprecision of models, this makes it possible to compare different upgrading options and optimise their preliminary design.

b) STUDY OF FILL MATERIALS

Because it is now forbidden in France to extract materials from the main channel of rivers, the sourcing of materials for building dike embankments is likely to diversify. As such, it is appropriate for the design stage of upgrading work to include an initial geotechnical study of embankment fill materials that can be sourced for a given site, whether they come from floodplain gravel pits or from quarries. Depending on the purpose of upgrading work (seal, mechanical reinforcement, shoulder at the slope base, etc.), it is possible to study the aptitude for compaction (see Appendix 1) of these materials, bearing in mind that a compacted embankment presents a number of advantages – improved seal and enhanced mechanical strength (greater density, increased resistance to shear and erosion, etc.).

c) TREATMENT OF PARTICULAR FEATURES

Particular points or features (pipe or tunnel crossings, constructions on the land-side slope, etc.) generally require specialist upgrading work, including making the transition with "standard" works carried out on either side.

d) EXECUTION DIFFICULTIES AND TEMPORARY WORK

It is quite common to underestimate the execution difficulties when at the preliminary design stage, especially in underwater work, such as the reinforcement of the toe of a dike in an area directly in contact with the river, or work where space is limited, such as on a land-side slope where there are buildings close by. Particularly when temporary excavations into the dike body are necessary, it should be seen to it that dike safety remains at least at its initial level (prior to work) and, if not, that the means to restore that level of safety prior to the next flood are available on the site.

The design office needs thus to be advised of such execution constraints and asked to take all temporary works and other provisions that may become necessary into account when evaluating different options of upgrading work – stop log structures, supports or temporary filling, special phasing of work, etc.

e) PHASING OF WORK

In general, the cost of upgrading per linear kilometre is high (over 760,000 euro per kilometre for Loire levees and 300 to 450,000 euro per kilometre for Camargue dikes). Obviously, because of budget considerations, there is a limit to the length of dike that can be consolidated each year. It is therefore good practice to draw up a programme of work to be done in different stages. The director of works (or operator of the structure) first needs to establish an order of priorities, taking into account how unsafe a dike is (failure probability) and its degree of vulnerability (the value of the human activities and interests directly protected by it) – see section 6.9. The design office could assist the director of works with this.

f) CONSIDERATION OF ENVIRONMENTAL CONSTRAINTS

The fact that a dike is a man-made structure should obviously not lead to dismissing environmentally-sound upgrading solutions. Naturally, some things cannot be entertained because potentially dangerous, such as encouraging or authorising growing trees or bushes on the dike body, base of an embankment and/or near masonry work.

Although it may not be possible to hide a dike, it is possible to make it more attractive by using traditional stone masonry or modern pre-fabricated parts that allow for herbaceous growth (e.g. perforated slabs on river banks).

Plant protection techniques can also be used for riverbank stabilisation work, provided there is no risk of woody vegetation growing on the body of the dike or near its base, which means they cannot be used in areas where the levee is in direct contact with the river. It is also advisable to verify whether such planting techniques suit the local hydraulic constraints and to ensure that regular maintenance of the protection structure can be carried out.

With the same concern to preserve the environment, the design stage could usefully aim at keeping the area to be worked on to a minimum. Whenever possible, preference could thus be given to (theoretically cheaper) upgrading solutions that involve only one side of the dike instead of both.

Finally, plans should be made for keeping negative impacts and detrimental consequences to a minimum when the work is in progress (installation of stop logs, watering of construction site tracks, etc.) and for anticipating any necessary compensatory and/or integration measures.

Worksite projects also provide the occasion for studying any means of access that may improve conditions for surveillance and maintenance work (e.g. service roads on dike crests and/or in front of the toe of both slopes).

g) FINAL TECHNICAL REPORT ON UPGRADING WORK CARRIED OUT

It is important to compile a detailed final report and a set of drawings following any upgrading work, if only because the work actually carried out nearly always differs from that planned, sometimes even down to the very nature of the work itself, but mostly in relation to geometric characteristics.

Final reports should include verified plans and cross-sections, together with a description of the work done, the difficulties encountered and the reasons for any departure from the original plan. An updated dossier on the structure is thus obtained, which also makes it possible to constitute a detailed chronology of any work carried out.

6.12 Assistance for the owner/director of works

We have just covered all the issues to be addressed in the process of in-depth dike diagnosis. As mentioned in the introduction, this stage should be preceded by a rapid diagnosis with clearing of vegetation, a topographical survey and an initial visual inspection.

Following this first stage and prior to in-depth diagnosis, we strongly recommend that a specialist be called in to identify the main risks to which the dike is exposed, so as to more accurately tailor the content of ensuing in-depth diagnosis. In some cases, this will result in a proposal to carry out very few studies.

For example, it serves no purpose to embark on onerous geotechnical explorations when a dike is wide in profile, has very gentle slopes and is built with known materials or, on the contrary, when its embankment is very weak and in a very poorly-maintained state (i.e. needing to be replaced in any case). On the other hand, a geomorphological study is very useful in the case of a dike built immediately next to the main channel.

The specialist's brief therefore basically consists of drawing up specifications for diagnostic studies and of helping the director of works to choose a design office and companies tasked with carrying out surveys. The cost will be largely offset by the savings generated by accurately defining the content of and time schedule for in-depth diagnostic studies.

The "Law on Water" dossier relating to modifications to dikes subject to French administrative authorisations could also be profitably entrusted to the same specialist.

BASIC TERMINOLOGY

(Refer also to figures 1 & 2)

"Banquette": Local French term used to describe a fuse plug structure built on the top of a spillway, comprising an erodible ridge of earth designed to wash out as soon as it is submerged.

Berm: Platform situated halfway up a dike slope, which generally allows access for surveillance and maintenance work.

Cemagref: French public agricultural and environmental research institute.

Crest wall gate or stop log assembly: Piece of wood (or sometimes metal), which slides into special grooves to seal off low areas on a dike crest or points giving access to a river or diked floodplain (gates, steps, spillways, etc.).

DDAF: "Département" Agriculture & Forestry planning office.

DDE: "Département" Town & Country planning office.

"Département": Administrative area (France is divided into nearly 100 of them).

Dike (dyke): Artificial structure built as a protection against flooding, at least part of which lies above the level of the natural terrain (locally and especially along the River Loire, the word "levee" is used, as it is in the USA) and which is designed to occasionally contain a flow of water in order to protect land liable to flooding.

DIREN: Regional branch of the Ministry of the Environment.

Downstream slope (of a dike): syn. with "land-side slope" (or "outer slope") of the dike (therefore on the opposite side of the dike from the river-side slope). The term "land-side slope" is preferable, to avoid confusion with "downstream" as applied to the course of the river – not relevant here.

EDF: French Electricity Board.

Foundation anchor block: Riprap abutment at the toe of a protective facing on the river-side slope of a dike.

"Franc-bord": Local French term (Loire valley) to describe the space between the main channel and the dike (its equivalent in the Camargue delta is *"ségonnal"*).

Freeboard ridge or device, freeboard masonry wall, berm: Structure built on the top of a dike body, comprising a ridge of earth (sometimes a masonry wall) and designed to provide protection against waves.

Gabion: Metal cage made of twisted mesh netting and filled with stones.

KM: Kilometre marker, used as a position-fixing reference and locating system, it gives the distance to or from a place (commonly used in dike, river, road and railways management). E.g. KM 1,750.

Land-side slope (of an earthfill dike): The dike slope that looks onto the flood-prone land that is protected by the dike (therefore on the opposite side of the dike from the river-side slope). Synonymous with "outer slope" of the dike.

Levee: synonymous with dike (dyke).

MM: Metre marker, used as a position-fixing reference and locating system, it gives the distance to or from a place. E.g. MM 1750.

Particular features: particular structures and features built into or across the dike (stop log structures and ramps providing access to the river, through-dike aqueducts, tunnels, culverts, conduits and pipes).

Protected area or [protected] valley: Land liable to flooding that is protected by a dike and situated between the dike and higher land (limit of the natural floodplain).

River-side slope (of an earthfill dike): The dike slope that looks onto a diked river or watercourse (therefore on the opposite side of the dike from the land-side slope). Synonymous with "inner slope" of the dike.

Ségonnal: Local French term used in the Camargue delta to describe the space between the main channel and the dike (synonymous with "franc-bord").

SN: Service de la Navigation

Spillway: Submersible, usually stone masonry structure designed to evacuate and spread flood waters onto outlying land in order, for example, to prevent dike overtopping.

Stone facing: Stone masonry facing on a river-side slope of a dike designed to protect the slope against water erosion.

Stop log assembly: See Crest wall gate.

Upstream slope (of a dike): syn. with "river-side slope" (or "inner slope") of the dike. The term "river-side slope" is preferable, to avoid confusion with "upstream" as applied to the course of the river – not relevant here.

Valley: Land liable to flooding and protected by the dike, lying between the dike and higher land (synonymous with protected area).

CHAPTER 1 – NATURE, FUNCTIONS AND COMPOSITION OF DIKES

1.3.1 – Fill dikes

Photo 1.1 – Typical view of a levee on the middle reaches of the Loire: fill dike with a broad cross-section, grassed slopes and crest roadway.

Photo 1.2 – Fill dike on the Petit Rhône in the Camargue (Bouches-du-Rhône), prior to upgrading work in 1994. Its crest is too narrow to allow plant and machinery to pass.

Photo 1.3 (courtesy of Frédéric Hédelin) – Aerial view of the diked downstream section of the Agly coastal river (Pyrénées Orientales), just after flooding on 12-13 November 1999.

1.3.2 – Masonry dikes and quay walls

Photo 1.4 – Stonework quay wall on the left bank of the Loire in Gien (Loiret), showing water level indicator and record flood levels.

Photo 1.5 – Old concrete protection on the bank and river-side slope of the Agly right-bank dike (Pyrénées Orientales). Recent repairs have been made in riprap.

Photo 1.6 – Work dating from 1958 on the Gardon river as it passes through the town of Alès (Gard): the bank is protected by concrete slabs that only locally serve as a dike.

1.3.3 – Spillways

Photo 1.7 – Masonry spillway at Ouzouer-sur-Loire (Loiret), with erodible earth ridge (grassed bank on the right of the photo) called "banquette" in France.

Photo 1.8 – Concrete spillway on the left-bank dike of the River Aude in Cuxac-d'Aude.

1.3.4 – Particular structures and features

Photo 1.9 – Construction embedded in the land-side slope of a River Cher levee at Savonnières (Indre-et-Loire).

Photo 1.10 – Closable sluice in a Loire levee at Val de Bou (Loiret), seen from the land side.

Photo 1.11 – Closable track on the right-bank dike of the Grand Rhône at Salins-de-Giraud (Bouches-du-Rhône).

CHAPTER 2 – CLASSIFICATION OF MALFUNCTIONS AND FAILURE MECHANISMS

2.1 – Overtopping

Photo 2.1 (courtesy of Frédéric Hédelin) – Aerial view of a breach caused by overtopping at St-Laurent-de-la-Salanque on the left-bank dike of the Agly, following flooding on 12-13 November 1999. The breach unfortunately appeared next to the waste-water treatment plant. Fortunately, the residents had been evacuated.

Photo 2.2 – Panoramic view (courtesy of Éric JOSSE, DDE 66) – Cross-section of the breach caused by overtopping at St-Laurent-de-la-Salanque on the left-bank dike of the Agly and damage to the waste-water treatment plant.

Photo 2.3 – Localised area of repaired overtopping on the land side of the left-bank dike of the Agly at St-Laurent-de-la-Salanque, following damage caused by flooding on 12-13 November 1999.

2.2 – External erosion and scouring

Photo 2.4 – Toe of a dike in contact with the main channel of the Loire on a concave bend of the Espagne levee at La Charité-sur-Loire (Nièvre).

Photo 2.5 – Foot of a scoured bank at the base of the Espagne levee at La Charité-sur-Loire.

Photo 2.6 (Source: ONF Alpes Maritimes) – Eroded bend caused by flooding of the River Var on 5 November 1994: the Guillaumes dike (Alpes Maritimes) with over 150 metres of the D2202 roadway swept away.

Photo 2.7 – River Agly dikes: foundation anchor block made of rip-rap at the foot of a bank partially swept away during flooding on 12-13 November 1999.

2.3 – Internal erosion

Photo 2.8 (Cemagref: Cyril Folton) – Damage caused by tunnel erosion (piping) in the new dike at Cuxac-d'Aude during flooding on 12-13 November 1999. A badly-installed conduit crossing was found to be the cause.

Photo 2.9 (Cemagref: Cyril Folton) – Downstream (land side) outlet of the piping.

Photo 2.10 (Cemagref: Cyril Folton) – Upstream (river side) outlet of the piping. Water was able to flow through the dike at a rate of several hundred litres per second. Breaching was only narrowly avoided.

Photo 2.11 (Source: Service Navigation Rhône-Saône) – Internal erosion (animal burrows or warrens, conduit crossings) was found to have caused all 16 breaches in the Camargue dikes during the two Rhône floods in the winter of 1993-1994.

Photo 2.12 (Source: Service Navigation Rhône-Saône) – Aerial view of a Camargue dike breach that coincides with a pipe crossing.

Photo 2.13 (Source: Service Navigation Rhône-Saône) – Leak on the downstream face of a Camargue dike caused by animal burrowing.

2.4 – Generalised slope failure

Photo 2.14 – Evidence of soil creep on a dike along the River Vidourle (Hérault). Notice the angle of the trees.

Photo 2.15 (Source: DDE 13) – Landslide following the retreat of flood waters – river-side slope of the left-bank dike of the Grand Rhône at Mas de la Ville.

Photo 2.16 (Source: DDE 13) – Embankment landslide affecting the crest of the structure (Camargue dike).

Photo 2.17 (Cemagref, Rémy Tourment) – Freeboard masonry wall on the river-side crest edge of the Val de Cisse levee (Indre-et-Loire). Access to the river in the background.

CHAPTER 3 – VISUAL INSPECTION OF DIKES & CHAPTER 4 – SURVEILLANCE DURING FLOODING

3.3 – Fill dikes

OVERTOPPING

Photo 3.1a (Source: DDE 66) – Damage caused by overtopping on the River Agly dikes following flooding on 12-13 November 1999. Extensive eroded zone in the embankment and high-water debris left on the crest.

Photo 3.1b (Source: DDE 66) – Detail of the eroded zone shown in the previous photo, revealing layers in the dike composition.

Photo 3.2 – Line of debris left by retreating flood waters on the river-side slope of the Cuxac-d'Aude dike after flooding on 12-13 November 1999.

Photo 3.3 – Post-flood inspection of the Cuxac-d'Aude dike: low points on the crest (passage of pathway) where overtopping occurred.

Photo 3.4 – Spillways need to be carefully inspected after flooding. Here, muddy traces left on the side wall make it possible to estimate the peak spillover height.

Photo 3.5 – Spillway at Sallèles d'Aude after flooding on 12-13 November 1999, showing damage to the gabion-built protective invert.

Scouring

Photo 3.6 – When old stone facing is exposed at the base of a river-side flood embankment, it may indicate undermining by the watercourse.

Photo 3.7 – River Agly dike inspection following flooding on 12-13 November 1999: exposed foundation anchor block made of rip rap, at the foot of a bank.

Photo 3.8 – Scouring at the foot of a protected bank on the Gardon d'Alès – gabion protection in poor condition.

Photo 3.9 – Sink hole and collapsed areas on the land-side top of the dike at Cuxac d'Aude after flooding on 12-13 November 1999. A conduit crossing was the cause of the damage.

Photo 3.10 – Inspection of the Petit Rhône dike near Arles after flooding on 8 January 1994: evidence of seepage across the embankment at the toe of the land-side slope.

Photo 3.11 – Sink holes along a dike crest are often a sign of internal erosion. Here, two sink holes appeared in the Agly dike crest within a few weeks of each other after the flood of 12-13 November 1999.

Photo 3.12 – Collapsed area on a masonry crest on a River Allier dike at Moulins (Allier), indicating internal erosion and/or settling of earthfill materials.

Photo 3.13 – The passage of pipes through embankments (here, a pumping station conduit drawing water from the Grand-Rhône) are points to be given particular attention during inspections.

Photo 3.14 – A non-return valve outlet at the toe of the river-side slope of a dike on the River Allier at Moulins.

Photo 3.15 – Section of the Val de Bou levee on the right bank of the River Loire, made narrower because of the presence of a house on the land side: a potentially vulnerable spot to be inspected during flooding.

GENERAL SLOPE INSTABILITY

Photo 3.16 – Escarpment and leaning trees on a river bank or on the river-side slope of a dike indicate general slope instability following the retreat of flood waters.

Photo 3.17 – Slopes with a gradient that is too steep (river side in this case) expose the dike to the risk of instability.

Photo 3.18 – Crack formation on the crest of a dike following the retreat of River Agly flood waters in November 1999 – sign of river-side slope instability following a rapid fall in the river water level.

MISCELLANEOUS

Photo 3.19 – Unstable freeboard masonry wall on the Authion valley levee (Maine-et-Loire)

Photo 3.20 – Dressed freestone (frequently used in the Loire valley) is subject to dissolution.

CHAPTER 5 – DIKE MAINTENANCE

Photo 5.1 – Dense ground cover (in this case, giant reeds) makes it very difficult to visually inspect this dike on the Petit-Rhône.

Photo 5.2 – Service tracks along dike crests (and in this case along the land-side toe of the embankment) facilitate surveillance operations on the River Agly.

Photo 5.3 – Tree roots distort, and sometimes break up, nearby stonework (Loire quayside in Tours).

Photo 5.4 – The breaches at Sallèles-d'Aude revealed the roots of plane trees running right the way through the base of the embankment.

Photo 5.5 – Animal burrows can cause serious damage if numerous and/or large and in places where embankments are narrow in section.

Chapter 6 – The basic principles of dike diagnosis

Photo 6.1 – Geophysical exploration using the radio magnetotelluric (RMT) method on the top of the Agly dike.

Photo 6.2 – Bathymetrical measurement using a weighted decametre on a transverse section of the River Vidourle.

Photo 6.3 – Sounding using a PANDA lightweight dynamic penetrometer on the dikes of the Vidourle.

Photo 6.4 – A PDG 1000 dynamic penetrometer set up for sounding.

Photo 6.5 – Sounding with a manual earth auger in the sandy-silt body of a Camargue dike.

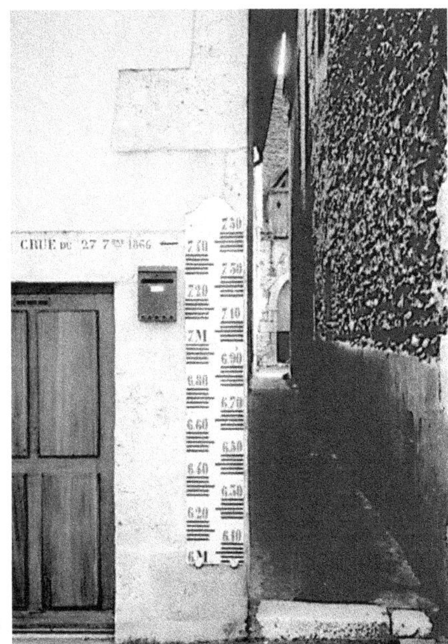

Photo 6.6 – Flood indicators on a house next to the Loire in Gien – one way of highlighting the risk!

Photo 6.7 – Widening of the land-side slope over a drainage blanket – Val de Bou levee.

Photo 6.8 – Emergency construction of a downstream drainage shoulder at the toe of the land-side slope of the Petit Rhône dike following the flood of 8 January 1994.

Photo 6.9 – Downstream drainage shoulder built on the Petit Rhône dike near Arles after flooding on 8 January 1994. Service track built into the top of the shoulder. The dike crest remains unsuitable for vehicles.

Photo 6.10 – Protection in the process of construction at the base of the river-side slope of the Val de Bou levee.

Photo 6.11 – Sheet piling as a protection for the base of the left-bank levee in St-Benoît (Loiret).

Photo 6.12 – Construction of a narrow diaphragm wall to create a seal in the body of a dike (Oder levee, Poland).

Photo 6.13 – River-side freeboard ridge and dike body strengthened with sheet piling on the Jargeau levee (Loiret).

Photo 6.14 – River-side stone facing repaired with pre-fabricated elements on the Authion Valley levee.

APPENDICES

APPENDIX 1
Basic principles of soil mechanics[1]

An earthfill dike is constructed from one or more soils. Most dikes, irrespective of their type, stand on a loose foundation formed from one or more layers of soil. The stability of the structure and of its foundation is therefore dependent on the mechanical properties of the soils.

A soil consists of three components:

– grains of material,

– water,

– and air.

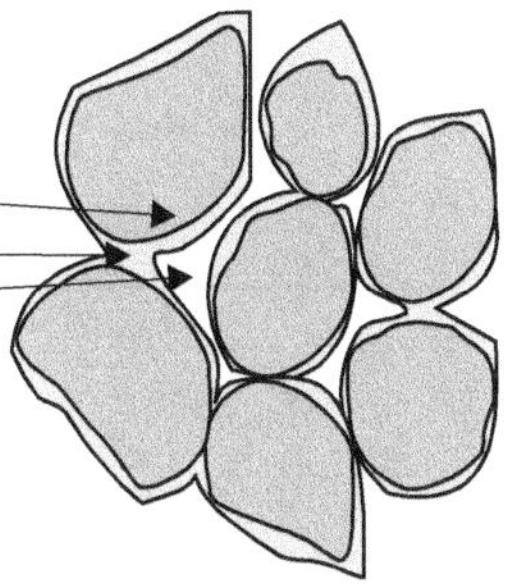

Figure 1. The physical structure of a soil

---

1. Taken from: G. Degoutte, P. Royet *Aide-mémoire de mécanique des sols*, Engref, Paris, 1999.

1. Characteristics of soils

The proportion of each of the three components indicated above varies depending on the context and nature of the soils under consideration. A loaded dike (one that is under pressure during a flood) will contain proportionally more water and less air than the same dike during a period of normal low water levels (known as a dry dike).

Equally, the clay used by a potter, or as a sealing element in a structure, is capable of retaining more water than sand or gravel.

The nature and origin of the grains of material that make up the soil can differ greatly. The properties of the soil depend to a large extent on the nature of these materials.

The nature of the grains in a soil can be characterized by their size: granulometry (the analysis of the grain size of the soils) differentiates several categories of materials, from very fine to very coarse.

Clays, loams or silts, sands, gravels and pebbles are some of the various classes of soil.

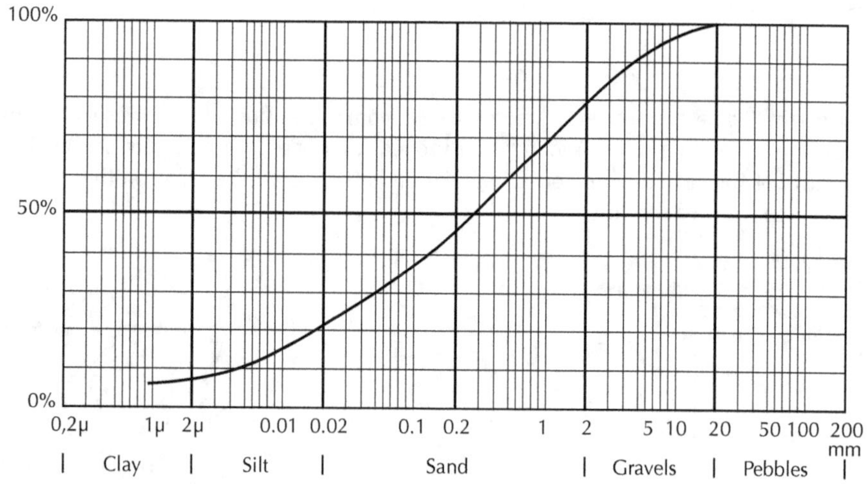

Figure 2. Granulometric curve

The grain size graph above shows the percentage, by weight, of grains that are smaller than the mesh diameter of the sieve. The graph is produced by sieving the soil to take out grains more than 80 μm in diameter. The sedimentometry test measures the proportion of the finest grains by decanting a soil suspension diluted in water.

2. Properties of fine soils

A "fine soil" is one for which 50% by weight of the grains are smaller than 80 µm. The behaviour of a fine soil is related to its water content, but also to its crystalline structure and its mineralogical composition.

Role of water: finer soils have a greater surface area per unit mass and consequently water, which is a polarised molecule, plays a proportionally greater role since it is a key factor in the electrical attraction forces between the grains.

The water adsorbed is the water attracted by polarity by clay particles. It forms a thin film around the grain, which can only be totally removed by very intense heating (200 to 300° C). This water, which is highly viscous, lubricates the grains.

The free water, which circulates freely between the grains and that can be removed in an oven at about 100° C, generates attractive forces due to capillary action.

In the laboratory, the role of water is usually evaluated by tests that measure the water contents at which changes in state occur: liquid to plastic to solid. These are known as the *Atterberg limits*:

$$w_L = \text{liquid limit}, \; w_P = \text{plastic limit}.$$

The plasticity index (PI) is a value $= w_L - w_P$.

A less common test measures the *shrinkage limit* (w_s), which differentiates a solid state without shrinkage, where the capillary water is present, from a solid state with shrinkage, where the capillary water has been partially removed. The removal of the capillary water causes a reduction in volume (shrinkage cracks).

This shrinkage phenomenon, whose sensitivity varies for different types of clay, is fundamentally important for the clay's sealing behaviour: the water-tightness of a hydraulic structure that is sealed by a clay-rich soil may be compromised by shrinkage cracks if it is does not come into contact with water for a long period, especially in a hot climate.

Solid state		Plastic state	Liquid state
with shrinkage $\quad w_s \quad$ without shrinkage		w_P	$w_L \qquad\quad$ **w**

Figure 3. Atterberg limits

The *plasticity index* PI $= w_L - w_P$ is extremely important in determining the conditions under which the materials are used.

Consequently, on earthfill construction sites, materials with a PI > 30 are difficult to use and compact.

3. Organic matter

Some soils produced by geologically-recent deposition may contain organic matter. They are identified *in situ* by their grey to black colour, the presence of plant debris and their odour.

In the laboratory, the overall content of organic matter is measured from the residue that passes through a 0.4 mm screen, dried at 65° C, and then reacted with oxygenated water. After a second period in the oven, the residue is weighed and the difference is the organic matter content.

If a soil with an organic matter content of more than 2 to 3% is used as a fill material, it can cause compaction problems in the long term. Soils containing more than 5% organic matter should not be used.

4. Soil compaction

Compacting soils can improve the mechanical and hydraulic characteristics of an earthfill structure. The energy provided by the compactor packs the soil grains closer together, increasing the density and strength of the mass. The air contained in the soil is expelled during this operation. Another effect of packing the grains closer together is to reduce the permeability of the fill material, thus improving the water-tightness.

For optimum compaction, the materials should be prepared. The layer to be compacted should not be too thick and the water content of the materials must be sufficient, without being excessive. In this situation, the water acts as a lubricant and facilitates the fitting together of the grains. However, an excess of water stops correct compaction: indeed, in this case, the energy provided by the compactor is absorbed by a layer of water that forms under the machine. This is known as the "mattressing" effect, which can be spectacular in extreme cases.

The Proctor test carried out in the laboratory determines the optimal density of a compacted soil at different water contents. A standardised amount of energy is provided. There is a Standard Proctor (SPO) test and a Modified Proctor (MPO) test.

When constructing large earthfill structures (dams, road embankments, dikes, etc.), the Standard Proctor test is used almost exclusively. In road building, the Modified Proctor is used, for which twice as much energy is provided.

The Proctor test generates a curve of dry weight by volume as a function of water content, for a given compaction energy. This test determines two quantities of fundamental importance to the operation and control of construction sites: the optimal water content (W_{OP}) and the optimal density (γd_{OP}).

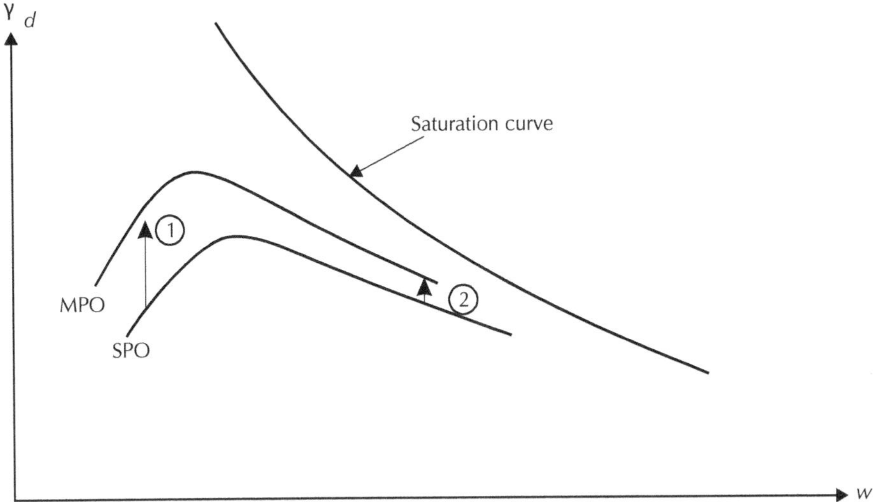

Figure 4. Proctor curves

On a compaction site, two quantities are controlled: water content and density, which must both lie within an acceptable range. For earthfills made from fine materials, the density value generally recommended is more than 98% of the optimum with a water content equal to the optimal water content + or – 1 or 2 percent (depending on the nature of the materials).

The Proctor curve is bounded by the so-called saturation curve. It represents the maximum density above which the material cannot be compacted any further (the limit curve for this soil condition).

Clearly, when water content is low, density can be increased by providing a greater compacting energy (1). However, when water content is high (> the optimal water content), there is little to be gained by increasing the energy (2). Worse still, some types of material react badly to over-compaction under these conditions and their density is reduced. Alternatively, saturated zones in the core of the fill will become the source of subsequent incidents.

5. Soil identification

Identification tests (grain size analysis, Proctor water content, Atterberg limits etc.) provide information about the nature of the material and its characteristics. These tests are carried out using reworked samples; they do not therefore provide information about the characteristics of the soil in the ground. If this information is required, there are various suitable '*in situ*' tests: static and dynamic penetrometer, pressure meter, vane shear tester, permeability test, plate test, etc.

Identification tests enable the soil to be categorised by name (clay, sand, clay loam, etc.). A naming system is very useful when the name refers to a strict definition, since a specialist in soil mechanic knows, for the type of soil specified, which properties should be studied and what the potential risks and main strengths are.

In summary:
– Clay or clay loam is a good material for producing the water-tight zone of a dike or earthfill dam.
– A sand cannot fulfil a sealing function.
– So long as it is sufficiently clean (i.e. with low proportion of fine particles), a coarse sand may be suited to the construction of a drain in an earthfill dike.
– A fine soil is more liable to settlement than a coarse soil.
– When constructing, a fine soil is more sensitive to water than a coarse soil.

6. Soil classification

In France, three types of classification are commonly used:
– The Casagrande diagram, which relates only to fine soils and uses Atterberg limits (Fig. 5).
– Taylor's triangular nomograph, which considers grain size only and is used to name a soil (clay, silt, sand, sandy clay loam, etc.) (Fig. 6).
– Standardised French classification NFP 11.300 September 1992 (Table 1), which replaces the former classification known as RTR (recommendations for earthworking for road applications) produced by the LCPC and SETRA, widely used in road geotechnology and which provides practical information about the ability of soils to be used as earthfill.

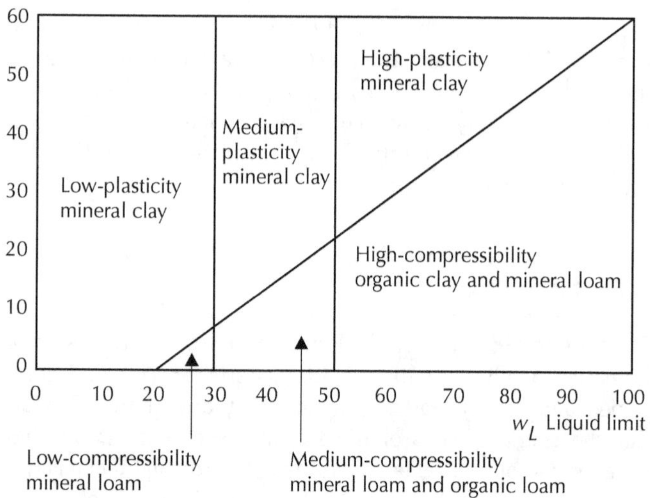

Figure 5. Casagrande plasticity nomograph

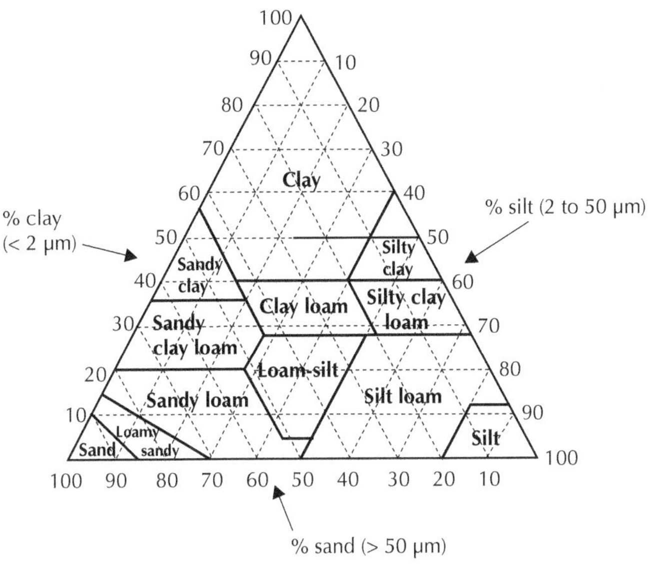

Figure 6. Triangular classification of fine soils

Table 1. Classification of soils (NFP 11-300) – simplified table

Fine soils **A** $D_{max} \leq 50$ mm $d_{35} < 0.08$ mm	$MBV^* \leq 2.5$ or $PI \leq 12$	**A$_1$**: low-plasticity loams, alluvial silts, fine clean sands, low-plasticity rock sand...
	$12 < PI \leq 25^{**}$ or $2.5 < MBV$	**A$_2$**: fine clay sands, loams, clays and low-plasticity marl, rock sand...
	$25 < PI \leq 40^{**}$ or $6 < MBV \leq 8$	**A$_3$**: clays and marly clays, high-plasticity loams...
	$PI > 40^{**}$ or $MBV > 8$	**A$_4$**: very plastic clays and marly clays
Sandy or **gravelly** **soils with fines** **B** $D_{max} \leq 50$ mm $d_{35} \geq 0.08$ mm	$d_{12} \geq 0.08$ mm $d_{70} < 2$ mm $0.1 \leq MBV \leq 0.2$	**B$_1$**: silty sands,...
	$d_{12} \geq 0.08$ mm $d_{70} < 2$ mm; $MBV > 0.2$	**B$_2$**: clay sands (low clay content)...
	$d_{12} \geq 0.08$ mm $d_{70} \geq 2$ mm $0.1 \leq MBV \leq 0.2$	**B$_3$**: silty gravel...
	$d_{12} \geq 0.08$ mm $d_{70} \geq 2$ mm; $MBV > 0.2$	**B$_4$**: clay gravel (low clay content)...
	$d_{12} < 0.08$ mm $\leq d_{35}$ $MBV \leq 1.5^{**}$ or $PI \leq 12$	**B$_5$**: sands and very silty gravel...
	$d_{12} < 0.08$ mm $\leq d_{35}$ $MBV > 1.5^{**}$ or $PI > 12$	**B$_6$**: sands and clayey to clay-rich gravels

Table 1. (*cont'd*)

Soils consisting of fines and large elements **C** $D_{max} > 50$ mm	$d_{12} < 0.08$ mm or $d_{12} > 0.08$ mm and $MBV > 0.1$	**C:** Flint clays, grindstone clay, scree, tills, coarse alluvium
Soils insensitive to water **D** $MBV \leq 0.1$ $d_{12} \geq 0.08$ mm	$d_{max} \leq 50$ mm $d_{70} < 2$ mm	**D$_1$:** clean alluvial sands, dune sand...
	$d_{max} \leq 50$ mm $d_{70} \geq 2$ mm	**D$_2$:** clean alluvial gravel, sands...
	$d_{max} > 50$ mm	**D$_3$:** coarse clean alluvial gravel, glacial deposits...

* MBV = methylene blue value (in grams of blue adsorbed per 100g of soil), characteristic of the clay content of the soil.

** An important parameter.

APPENDIX 2
Basic principles of soil hydraulics[2]

The hydraulic properties of soils determine the ability of structures and their foundations to be watertight, to resist internal erosion (piping) and to maintain their component materials in place (no clogging of the drains).

1. Preliminary definitions

1.1 Velocity of water in soil
By definition, the apparent velocity v is equal to the unit quantity of flow of water (Q) that flows across a unit cross-section of the soil sample (S), i.e. $v = Q/S$. In fact, the actual average velocity (between the grains) is v/n where n is the porosity, but an acceptable simplification is to work with the apparent velocity. In the following analysis, v will always be the apparent velocity.

1.2 Hydraulic head at a given point
Consider a point located in a saturated mass of soil through which there is a continuous flow. Let u be the water pressure at this point and z its height above a datum. The hydraulic head at this point is, by definition:

$$h = u/\gamma_w + z + v^2/2g$$

where γ_w: is the density of water.

The velocities in the soil are always small, and so the $v^2/2g$ term can be ignored.

Hence: $h \approx u/\gamma_w + z$

1.3 Hydraulic gradient
In a uniform and unidirectional flow, the hydraulic gradient i is, by definition, the difference in hydraulic head Δh divided by the length L of the path of the water through the soil.

---

2. Taken from: G. Degoutte, P. Royet *Aide-mémoire de mécanique des sols*, ENGREF Paris 1999.

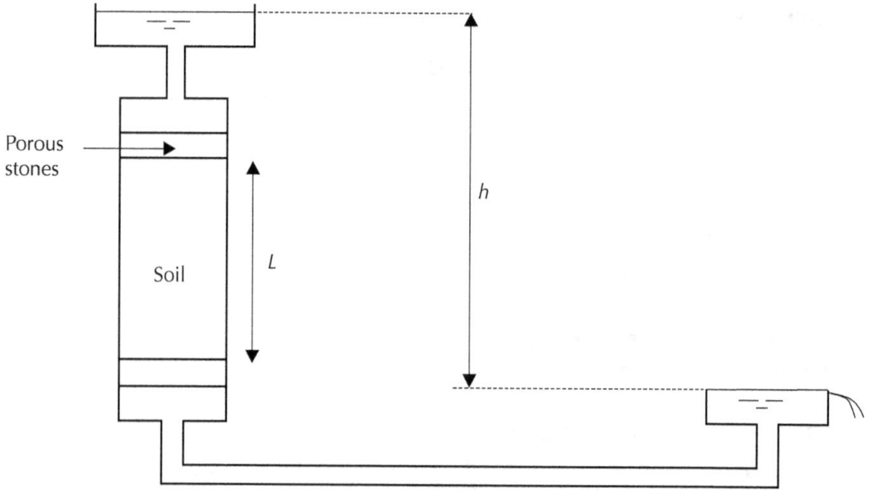

Figure 1. Principle of the permeameter: a soil sample subjected to a hydraulic gradient $i = h/L$

2. Hydraulic properties of soils

2.1 Darcy's law

This fundamental relationship is written as $v = k\,i$ (where v is the water velocity, i the hydraulic gradient and k is the coefficient of permeability of the soil). This coefficient is equal to about 10^{-8} to 10^{-10} m/s for a clay and 10^{-4} to 10^{-6} m/s for a sand.

2.2 Equipotentials and flow lines

Equipotentials are lines along which the head h is constant. They are orthogonal to the flow lines.

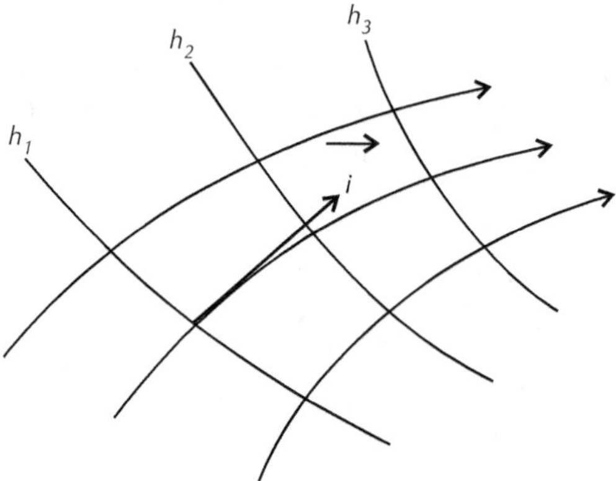

Figure 2. Flow net of flow lines and equipotentials

Case study: a dam or dike with a horizontal drain on an impermeable substratum

Consider a dam or dike that is drained horizontally, in a context of continuous flow. There are two boundary conditions in steady hydraulic state: $h = H$ along the length of the dam face AB and $h = z$ along BC (saturation curve).

The head is zero at the drain level. The saturation curve and the contact with the foundation are flow lines. Hence the appearance of the plot in Fig. 3.

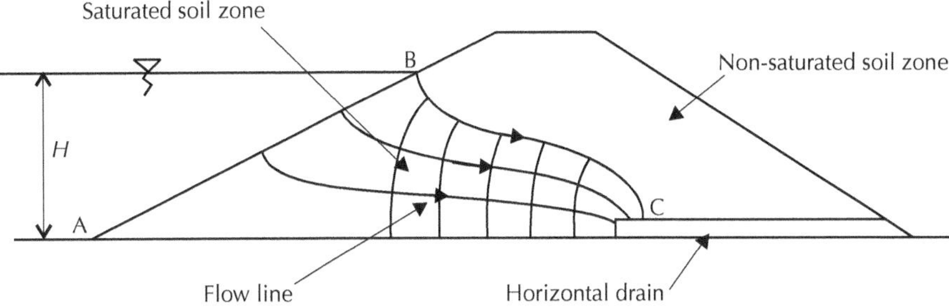

Figure 3. Saturation of a hydraulic earthfill structure

For reasons of simplification, the example illustrates a horizontal drain in an isotropic medium. However, the practicalities of earthfill construction (using compacted layers) usually result in a high degree of anisotropy in the soil, with much higher permeabilities in the horizontal direction.

The total seepage flow rate can be determined by summing the leakage flow rates in each current tube (calculated using Darcy's law).

2.3 Flow forces and gravitational forces in a saturated soil

The force of gravity γ' applied to a grain of soil is proportional to its mass and acts in the vertical direction.

The flow force $\gamma_w i$ acts tangentially to the flow line, and is proportional to the hydraulic gradient i.

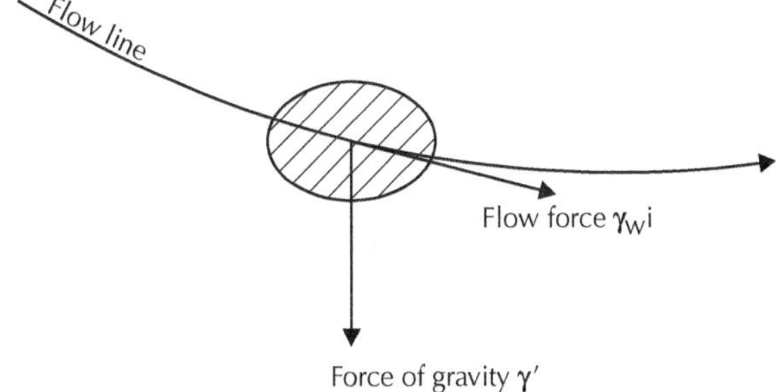

Figure 4. Forces applied to a grain of soil

2.4 Critical gradient; piping

The piping phenomenon occurs when the flow force is ascending and greater than the force of gravity, i.e. $\gamma_w \cdot i > \gamma'$. Hence the critical gradient: $i_c = \gamma'/\gamma_w$. For a sand with a void ratio e = 0.7:

$$\gamma' = \frac{\gamma_s - \gamma_w}{1 + e} \approx \frac{27 - 10}{1,7} = 10 \text{ kN/m}^3,$$

where γ_s: the density of the grains = 27 kN/m³.

The critical gradient i_c is thus equal in this case to 1. In Figure 5 below, $i = h/(2.L)$ and the piping appears when the drop in water level in the chamber reaches $h = 2.L$.

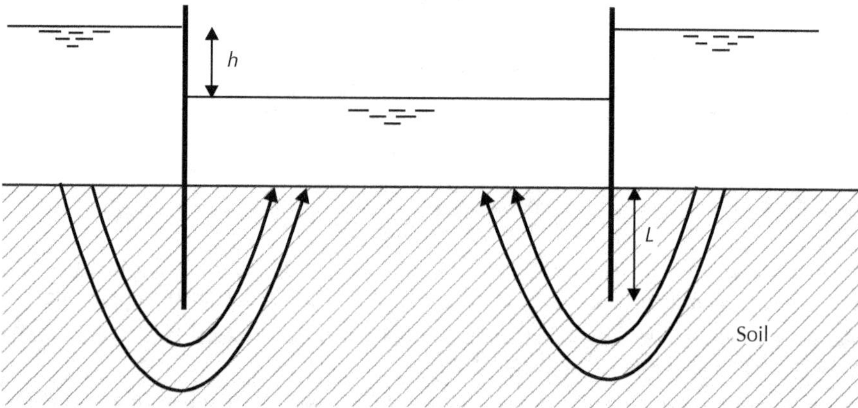

Figure 5. Flow under a chamber created from sheet piles

For a loaded dike, hydraulic piping appears preferentially at the toe of the structure (where the hydraulic gradient is highest) or along lines of preferential flow generated by heterogeneities (traversing pipes or conduits, burrows or roots).

2.5 Measuring permeability

• *In the laboratory*, the permeability is measured using a permeameter (the principle is illustrated in Fig. 1):
– Head-constant mode: the flow rate required to keep the upper reservoir full is measured.
– Variable-head mode: the drop in the water level in the tube (under the upper reservoir) is measured as a function of time.

• When testing *in situ*, the standard test (standard NFP 94-130) is to pump water into a borehole, at a constant flow rate Q until steady hydraulic state is reached (h = constant). It has been demonstrated that, during the groundwater drawdown test in steady hydraulic state, the coefficient of permeability is obtained from the equation:

$$k = Q \frac{\ln(R/r)}{\pi(H^2 - h^2)}$$

with the heights h and H measured with respect to the impermeable substratum.

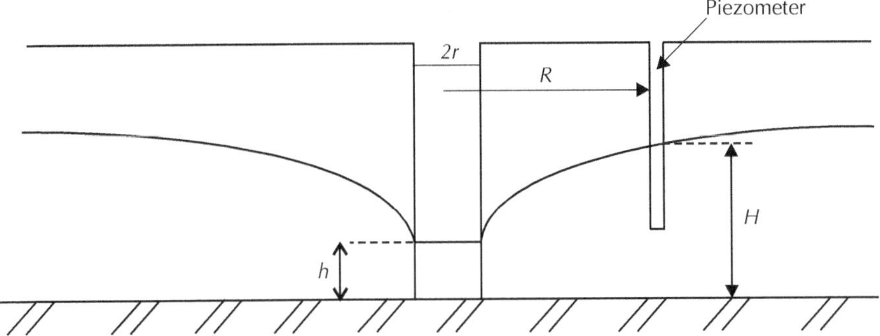

Figure 6. Groundwater drawdown

In permeable soils under the water table, the Lefranc test is also carried out (standard NFP 94-132): pumping or injecting water at a constant flow rate into a borehole and measuring the variation in head with time.

In rocky masses, the Lugeon test is performed (standard NFP 94-131): injecting water at constant pressure into a borehole.

3. Filter rules

Carried by a flow of water, fine soil particles will, in some cases, migrate towards a coarser zone of soil. A good example of this is the process that occurs between the earth-filled body of a hydraulic structure and the drain material. To prevent this phenomenon, two successive zones of a hydraulic structure must satisfy the **filter conditions,** which are rules relating to grain size transitions. In practice, if the filter conditions are not satisfied between two adjoining materials, then a material with an intermediate grain size distribution should be included between them, called a filter. The conditions described below must be satisfied at the two interfaces: between the fine material of the fill and the filter, and then between the filter and drain. In each case, D_a and d_b are the diameters of the sieves that allow $a\%$ by weight of the coarsest material D (drain) and $b\%$ by weight of the finest material d (fill) to pass through.

• When a fine material with a *continuous* grain size distribution (no change in the gradient of the granulometric curve) is in contact, in a hydraulic structure, with a uniform material (drain or filter), their grain size distributions must satisfy the following conditions:
– Condition for non-entrainment of fines: $D_{15} < 5.d_{85}$.
– Permeability condition: $D_{15} > 0.1$ mm.
– The coefficient of uniformity (D_{60}/D_{10}) of the filters and drains must be between 2 and 8.

Usually, a cleanliness condition (i.e. low proportion of fine particles) is also required for the material that constitutes the drain, a condition that can be written, for example, as: $D_{05} > 0.08$ mm.

- The filter condition for the contact between two highly uniform materials ($D_{60}/D_{10} < 3$ and $d_{60}/d_{10} < 3$), – which is the case between the filter and drain – is written as: $5d_{50} < D_{50} < 10d_{50}$.

- Finally, a highly-graded soil ($d_{60}/d_{10} > 16$) with a discontinuous grain size distribution is very likely to have its fine particles carried away by internal erosion due to the effect of flows of water. The filter abutting this material must therefore be determined using the d_{85} of the lower part of the granulometric curve for the soil, after the change in gradient of the curve (as shown in fig. 7).

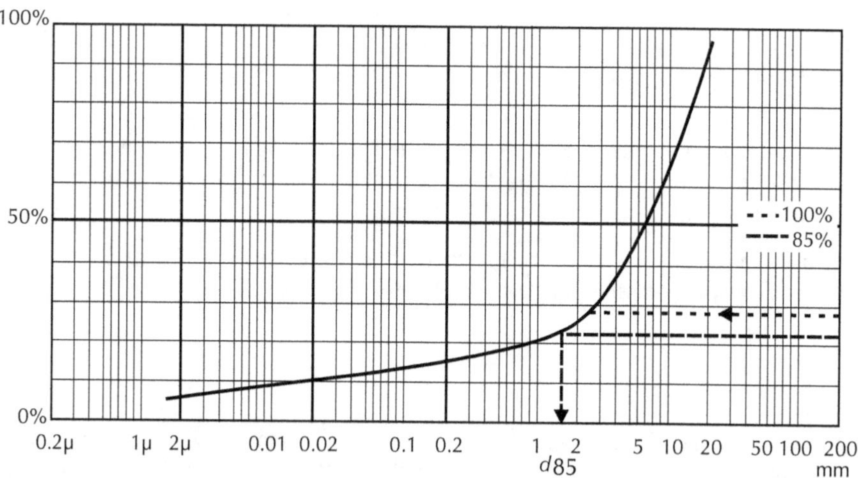

Figure 7. Case of a soil with a discontinuous grain size distribution

Case study: sizing the grain size transition of the protection of a bank or dike

Consider a bank or a sandy dike, whose granulometric curve is shown below, and a protection against scouring provided by 300-700 mm blocks (i.e. weighing 50-400 kg).

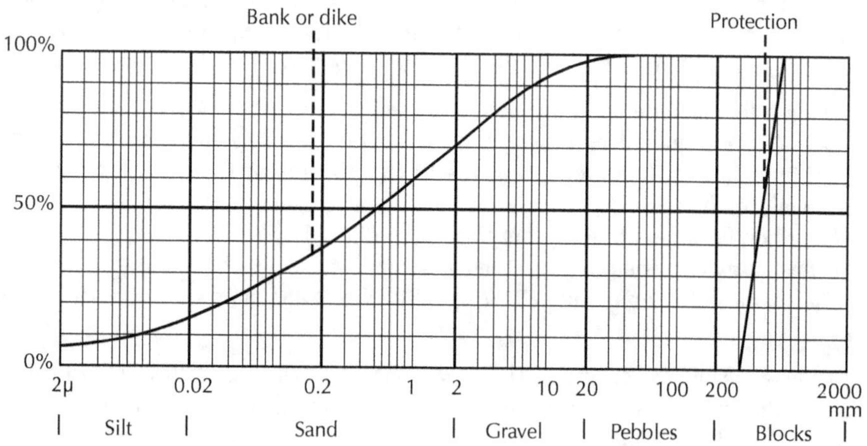

1. Need for a transition

A transition is required if $0.2.d_{15}$ (protection) $> d_{85}$ (bank or dike).
In this case:
d_{85} (bank or dike) = 5 mm and d_{15} (protection) = 350 mm. 0.2x350 = 70 > 5.
The condition is not even remotely satisfied. A transition is therefore required.

2. D_{50} condition for the transition

Since the transition and protection are constant, the condition to be respected is:
$5.d_{50}$ (transition) $< d_{50}$ (protection) $< 10.d_{50}$ (transition).
Hence $5.d_{50}$ (transition) < 450 mm $< 10.d_{50}$ (transition),
that is 45 mm $< d_{50}$ (transition) < 90 mm (segment AB in the graph below).

3. D_{15} condition for the transition

0.1 mm $< d_{15}$ (transition) $< 5.d_{85}$ (bank or dike)
That is 0.1 mm $< d_{15}$ (transition) < 5x5 = 25 mm (CD segment).
The grading zone shown below is therefore appropriate since, moreover $d_{60}/d_{10} = 6$ or
7 falls between 2 and 8.

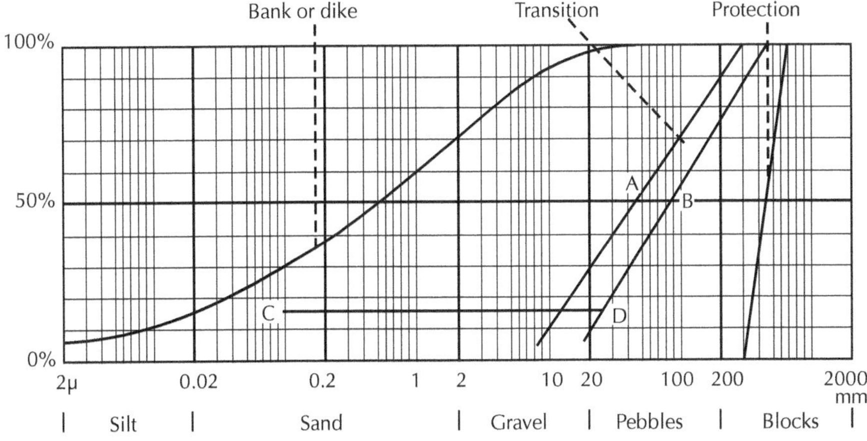

Appendix 3
Surveillance dossier for 'dry' dikes
How to perform a visual inspection of a fill dike:
– Initial inspection
– Routine surveillance

1. Principle and objective of the inspection

The inspection should consider the entire length of the dike, recording all visual information about both the external *morphological characteristics* of the structure and the existing or suspected *anomalies* that affect the component elements of the dike.

The initial inspection is an essential step in any diagnosis (rapid or in-depth) of a dike system.

Routine surveillance is one stage in an ongoing surveillance process, for which an initial inspection is an essential prerequisite. Routine surveillance detects any changes that have occurred in the dike and the surrounding area since the previous inspection.

1.1 Morphological characteristics of the dike

The extent of the basic topometric survey work that can be performed during the visual inspection is clearly dependent on the quality of the topographical documents available beforehand about the structure and its surroundings:

(A) WHERE AN UP-TO-DATE AND ACCURATE TOPOGRAPHICAL MAP IS AVAILABLE (WITH A SCALE OF ABOUT 1:500):
This is the ideal situation for a correct diagnosis of the structures. The topographical map provides an invaluable support onto which information collected during routine surveillance can be copied.
The 'in situ' work then simply involves checking and completing the principal topographic information available, which requires tracking your position on the existing map as the inspection progresses. Transverse profiles are only surveyed at those cross-sections that contain particular features that are invisible or incorrectly described on the map (e.g. a house or building built near to the dike or embedded in the slope).

(B) WHERE ONLY A 1:25000 NATIONAL SURVEY MAP IS AVAILABLE:
In this case, the following information may be quite easily recorded without making the visual inspection too time- and material-consuming:
– Width and camber of the crest, space taken up by any roadway.

- Gradient and length of the river-side slope, distance from the main channel to the toe of the dike.
- Gradient and length of the land-side slope.
- Water level(s) on the day of the inspection, peak flood level and/or high-water marks.
- Mention of any structures or constructions and of any particular topographical features (e.g. a depression on the land side).

The most efficient way of recording this information is to record successive transverse profiles using a decametre and a clinometer (a simple optical device for measuring slope gradients, the same size as a pocket compass). The profile should extend significantly beyond the toe of the slope and be linked to an identifiable point on the 1:25000 scale national survey map, which is available in most countries. Particular features, such as constructions or indicators of anomalies, are identified on the transverse profile and can then be correctly positioned when the data is formally drawn up in the office. The longitudinal positioning of the profiles may be determined using a Topofil (a distance measuring device based on unreeling a thread which is not retrieved) or a GPS receiver (if such a device can be used), in combination with position fixing using the dike's KM system.

1.2 Indicators of anomaly

The main elements to be examined are listed in the tables in the main text of this handbook (see Tables 1 and 2 in section 3.3). The elements to be examined are categorised on the basis of failure mechanism and the part of the structure being investigated (slope, crest, etc.).

Photographs of the most important anomalies are helpful asset and can be used to make visual comparisons with photos taken subsequently from the same viewpoint (in the context of routine surveillance). For best results: use the flash if facing into the light or if the light is poor; place an object next to the subject to give an idea of the scale; identify and record the location and the angle of the shot.

2. General procedure

2.1 Preparing for the inspection

Prior to the initial inspection, it is essential to collect and analyse all the documents available that relate to the dike: current and out-of-date topographical maps, drawings of gated structures, study reports, work reports, historical documents (complaints from local residents, damage reports, description of breaches, etc.).

Prior to a routine surveillance exercise, the first step is to analyse in detail the documents produced after previous inspections.

If a detailed topographical map is available, then exact preparations can be made that will simplify subsequent operations in the field:
- Choice of datum kilometre marker (KM).

– Determination of the descriptive segments (see section 2.2 below).
– Noting on the map the boundaries between segments and listing possible repositioning points in the field.
– Preliminary list of particular features identified on the map (constructions, walls, roads, bellmouth structures, large isolated trees, gates, ponds, inspection holes, etc.).
– Filling in the basic information on the inspection reports (see section 3 below): KM positioning, length of each segment, name of the locality, etc.

The equipment checklist for the visual inspection is:
– a set of 1:25000 national survey maps and any available detailed maps.
– a Topofil device (or a 50-metre tape measure) and/or a GPS pocket receiver.
– a clinometer and a pocket compass.
– a machete, a marker and an aerosol can of paint.
– a geologist's hammer, a folding U.S. shovel, a felt-tip pen and sample-collection bags.
– a pocket ruler and a tape measure (50 metres or 20 metres).
– a 24 x 36 printed photo reflex camera or a digital camera.
– a sketch pad with pencils and rubbers.
– a set of blank forms for describing the anomalies and/or the transverse profiles.
– for a routine inspection, a set of forms describing the anomalies and/or the transverse profiles recorded during the previous inspection.
– several marker pegs.
– safety equipment.
– as an option: a manual earth auger.
– as an option (if working on an electronic version of the form): a laptop computer.

For optimal conditions of observation, the ideal season to work in is winter, after, if possible, clearing growth off the slopes of the dike. If necessary (when the dike lies directly alongside the main channel), an additional inspection of the submerged toe of the slope and/or of the banks should be scheduled for a low-water period and/or from a small boat.

2.2 Performing the inspection

The field team is made up of two or three people trained in civil engineering or soil mechanics: working in pairs as a minimum has been found to be essential in terms of carrying the equipment comfortably, recording the geometrical characteristics quickly and for safety reasons. Working as a group of three is worthwhile for reasons of efficiency when the levee to be surveyed is high and wide and/or poorly maintained, or where there is no detailed topographic information (and where there are numerous transverse profiles to survey). The inclusion of a technician qualified in civil engineering in the team is recommended to ensure a more exhaustive and detailed listing of any anomalies and to ensure that subjective assessments of risks affecting the long-term future of the dike are balanced by a viewpoint from a relevant discipline.

It is advisable to inspect the anomalies by following a methodical route over the dike. One simple way of working is to divide the dike into segments of a length that is predetermined (and adapted to the complexity of the structure):

100 m long for well-maintained zones and 20-25 m long for segments covered with vegetation or badly degraded (numerous anomalies and particular features). If A and B are the two ends of the section to be investigated, one possible route for the inspection team is to walk from A → B along the crest of the dike, laying down markers (marker pegs and Topofil thread or decametre), then to return from B → A along one of the slopes (without forgetting to examine the toe of the slope and the bank of the river, if it is nearby) and, finally, a second transit from A → B to cover the other slope.

Any local residents encountered by chance during the visit should be asked questions about how the dike operates and about any recent maintenance work carried out. Their observations are noted in the comments boxes of the anomalies record form or on the transverse profiles plot.

Recording information in the field:

The data logging method will vary depending on the quality of the maps available:

A) WHERE AN UP-TO-DATE AND ACCURATE TOPOGRAPHICAL MAP IS AVAILABLE (WITH A SCALE OF ABOUT 1:500):
In this case, it is possible to work in the field directly on a copy of this map (with reference made to the anomalies record forms – see the example in part 3 below – for detailed comments and transverse profiles recorded at particular sections). This information can then be compiled properly back in the office at a later date. The forms may be completed in the field using a portable computer.

B) WHERE ONLY A 1:25000 NATIONAL SURVEY MAP IS AVAILABLE:
This situation should only occur if there is not enough time or resources to carry out the initial topographical work that we recommend. The description in the field will then be based on recording transverse sections with respect to a systematic basic survey grid (e.g. one profile every 100 or 50 metres), with any particular points of interest between profiles identified in terms of their KM and entered in the form (see the blank form in the appendix). Additional profiles may be surveyed at sections of interest (e.g. a structure or a construction on or in the body of the dike, steepening of a slope, etc.). A fair copy of all profiles is then written once back in the office and, as a minimum, they are to be identified in terms of their KM on the best cartographic medium available.

2.3 Writing up the inspection back in the office
This work mainly consists of collating and writing up fair copies of the information collected in the field: notes, sketches and transverse profiles. Time should also be allocated for filing photographs and for writing detailed captions.

Entering the data onto a computer is valuable for the detailed surveillance of large structures: simple software such as standard spreadsheets or databases is more than sufficient for this purpose. The ideal solution would be to develop a specially-written computer application that would give structure to and homogenise the data entry process for a sufficiently large group of users. Entering data is always time-consuming;

however, savings can be made by ensuring that fair copies of data are written in the office by one of the technicians that took part in the inspection.

3. Example of the use of a typical anomalies record form

3.1 Foreword

This example assumes that an up-to-date 1:500 or 1:1000 topographical map is available of the sector of the dike to be inspected.

The inspection is made on foot of elementary segments, of predetermined length, whose position is referenced to an existing KM system and indicated temporarily on the ground by at least one of the operators (Topofil or tape measure and marker pegs).

The surface of the dike to be described is divided up into four parts:
– Bank of the river.
– Slope and toe of the dike on the river side.
– Crest (including any freeboard feature).
– Slope and toe of the dike on the land side.

For each part of the dike in a segment, the anomalies and information about the composition of the structure are marked on the map (to scale wherever possible) as they are observed, with a reference corresponding to a line on the record form.

The aim of the form is to provide, as a complement to any sketches or drawings and captions written directly on the map, coded and alphanumerical information (with comments) that may be stored in a database and processed by data analysis tools, in a shared file format, e.g. to allow it to be accessed by a group of managers.

3.2 How to use the form

General layout of the form:
The form is used to describe the four parts of an elementary segment of a dike: bank, river-side dike slope, crest, land-side dike slope. The two boxes that appear at the top of the form provide general and geographical information about the elementary segment. The body of the form is divided into 4 main boxes that relate to the 4 parts of the dike defined above. The boxes at the bottom of the form provide a reminder of the coding used.

Header information:
This information should be filled in before starting work in the field, with the exception of the "Date" and "Operators" fields.

• **Date:** the date of the inspection.

• **District:** name of the district in which the elemental segment is located.

• **Precise location:** name of the village or hamlet closest to the segment, indicated on the 1:25000 scale map.

• **Operators:** names of the people carrying out the work, starting with the name of the person filling in the form.

• **Segment_length:** length (in metres) of the segment described. This corresponds, other than in unusual cases, to a fixed length of elemental segment that is selected beforehand in the office for the entire portion of the dike to be inspected. If, during a brief prior inspection, the dike appears to be extremely heterogeneous, badly cleared (insufficient visibility due to the vegetation) and/or appears to have numerous anomalies, a reduced segment length can be chosen (25 to 50 m).

• **Ref. of the KM:** reference in plain text of the basic KM used for the longitudinal reference system.

• **Bank:** RB (right bank) or LB (left bank), as appropriate.

• **Page:** to identify a page number if the description extends onto several forms.

GENERAL POSITIONING BOX (FOR THE SEGMENT DESCRIBED):

• **KM:** kilometre marker coordinates for the *segment starting point,* in relation to the datum KM used for longitudinal positioning. This field is, if possible, filled in beforehand in the office when the elemental segments are defined. Unless otherwise indicated, the KM for segment "n" is equal to the KM of segment "n-1" plus the length (Segment_length) of the same segment "n-1".

• **MM_sta:** indication, in terms of metres, of the *starting point of the segment,* provided by the ground-measuring device (Topofil thread or tape measure).

• **MM_end:** indication, in terms of metres, of the *end point of the segment,* provided by the ground-measuring device (Topofil thread or tape measure).

In principle, MM_end = MM_sta + Segment_length. However, if a repositioning point with respect to the basic KM is used at some point along the length of the described segment, then the metric end point is displaced, if necessary, so that it matches up exactly with the KM positioning. Finally, if the measuring device on the ground (Topofil thread) is not reset, the MM_sta of segment "n" is taken to be equal to the MM_end of segment "n-1".

BOXES FOR DESCRIBING ANOMALIES AND OBSERVATIONS FOR EACH PART OF THE DIKE:

The body of the form is divided up into 4 main boxes, relating to the 4 parts of the dike: river bank, river-side slope and toe, crest (including any freeboard feature or elevation), and land-side slope and toe. The general structure of the boxes is identical for each part of the dike: only the "anomaly" and "structure element" codes may differ (e.g.: the "flood level" [FLV] code is only applicable to the two parts on the river side (bank or dike slope), and the structure element "freeboard feature" [FBF] only exists for the crest).

< Identification/description of the anomalies >

For each of the 4 parts of the dike, there are, in the left-hand 3/4 of the correspon-ding box, 6 or 8 lines, numbered 1x to 8x (or 6x) (where "x" is a symbol specific to each part and designed to prevent any confusion when written on a map: "b" for bank, "r" for river-side slope, "c" for crest, "l" for land-side slope. These lines are used to enter all the (indicators of) anomalies or particular features:

• **Anom. Ref.:** reference, between 1x and 8x (or 6x), that refers to the same number written on the 1:1000 scale map, in the dike section concerned. If there are more than 8 (or 6) anomalies for any one part of the segment, a second form should be used (not numbered), with the line numbering resumed at 9x (or 7x) and indica-ting page-numbering information in the top right-hand corner of the form. On the map itself, the information should, if possible, be symbolised (using a standardised caption) and/or drawn to scale.

• **Anom. code:** a 3-character alphanumeric code that describes the nature of the anomaly (or indicator or particular feature). This code refers to a key at the bottom of the page. Some codes are only applicable to a specific part of the dike (e.g.: **FLV/WLV** for a flood level/water level, which can only be used on the river side of the dike). Other "anomaly" codes are more specifically concerned with rigid, masonry structures (made from cut stone or concrete) e.g.: **LOO** for loosening, **LSC** for loss of structural cohesion, etc. – see the corresponding table). Four special codes are also available to identify actions taken during the inspection: **PZO** for a piezometer reading (if possible, the level measured during the visit should be recorded), **SDG** for a sounding, **SAM** for a sampling of material from the dike (indicate the depth and a reference if applicable) and **TPF** for a transverse profile (with reason given for why it was surveyed).

Struc. code: a 3-character alphanumeric code used to indicate, if necessary, the dike structure affected by the described anomaly. E.g. the "structure" code [**PTF**] combi-ned with the "anomaly" code [**CRK**] indicates a crack on the protective facing.

• **Nb:** number of anomalies found for which this reference number applies. If the number is greater than 1, it is a set of anomalies (e.g. animal burrows), or is grouped within a limited zone (**Long. pos.** $\neq$ empty), or distributed over the entire length of the segment (**Long. pos.** = empty).

• **Long. pos.:** a value in metres indicating the longitudinal positioning "x (m)" of a particular anomaly: an absolute value between 0 (minimum) and the value of MM_end – MM_sta (maximum) or a pair of ends of the section "x1 (m) – x2 (m)" used to identify one or more anomalies spread over a length of several metres. If the field is left empty, then this indicates that the anomaly extends over the entire length of the segment.

• **KM (calculated):** kilometre marker used to identify the anomaly, based on the KM reference system, calculated in the office or by the computing system (if **Long. pos.** $\neq$ empty), taking into consideration any correction of the metric positioning resulting from KM repositioning.

For example, if an error in measuring the longitudinal positioning may be entirely attributed to the last segment profiled, the KM of the anomaly is calculated as follows (with Long. pos. > 0):
KM[anomaly] = KM[segment] + Long. pos.[anomaly] / 1000 * (1 + (MM_end – MM_sta – Segment_length) / Segment_length))

If this is not the case, the difference found must be distributed over several successive segments by applying an appropriate corrective equation.

• **Photo** fields**:**
In the **Nb** column the number of (detailed) photos taken of the referenced anomaly is indicated and, in the **ID** column, the photo identification number(s), is indicated in the form "nb1-nb2" if there are several photos.

• **Description of the anomaly:** a free comments zone that is filled in with any information that might specify the nature, extent, location or characteristics of the anomaly.

sev. code: a subjective scoring of the severity of the anomaly, assessed *locally* (i.e. in terms of the safety of the dike element affected, and not that of the entire dike):
=> code 1: start of an anomaly, insignificant and/or infrequent anomaly
=> code 2: significant and/or fairly frequent anomaly
=> code 3: very significant and/or omnipresent anomaly

E.g. a score of 3 attributed to animal burrows [BUR] distributed over the entirety of a slope [SLP] could indicate a very high density of burrow openings.

< Observations for all (or part) of the segment >
The right-hand side of the description box for each of the 4 parts of the dike is reserved for information that relates to the entire segment, for the part concerned.

The first 3 or 4 lines (depending on the case) are used to indicate the presence or not (**Y/N/U** code for Yes/No/Uncertain) of ancillary longitudinal structures. The list proposed includes the main types of structure that may be encountered on a specific part of the dike: PTF for protective facing on the river-side slope, FBF for a freeboard feature on the crest, etc. Additional information can then be entered (in the **Details about the nature of the structure** field) about the nature and location of existing features (e.g. whether or not they cover the entire length of the segment).

The **Access** field is used by inserting **Y/N/U** to indicate if there is access for construction vehicles at the toe of the dike (for both parts on the slope) or on the crest, depending on the individual case.

The **Overall view photo** is used to reference a general shot of the part of the dike concerned:
– **Photo pos. (m):** value in metres (Topofil thread or tape measure) of the point from which the photograph was taken.
– **Upstream <—> Downstream:** cross out the symbol not required (< or >) to indicate the direction in which the photo was taken (upstream => downstream or downstream => upstream).
– **Number:** the photo number, read off the camera.

The **Comments** zone is used to make general remarks about a part of the segment, or about the entire section, and to mention any specific points: e.g. information provided by local residents and their contact details, etc.

3.3 Tables of anomaly codes and glossary

ANOMALY CODES – ALL PARTS OF THE DIKE:

• **BUR:** BURrow or tunnels of burrowing animals.

• **CON:** CONduit or pipe outflow, through pipes, box culvert or ancillary structures (e.g. inspection hole).

• **CRK:** CRacK in the ground or in a rigid structure.

• **GER:** Gully ERosion indicator on the slope or platform (in principle in the transverse direction).

• **MLE:** Miscellaneous Longitudinal Erosion, other than that caused by the watercourse (e.g. a step in the toe of the dike created alongside a path or platform).

• **MVT:** settlement, sliding, soil creep or any other indicator of the MoVemenT of earth or of a rigid structure (including the tilting of a wall or of sheet piling).

• **PAS:** PArticular Structure other than a pipe or conduit (e.g. a construction, cellar, wall in the body of the dike, opening/gate in the freeboard wall).

• **SIN:** SINk hole – a surface sign of piping in the dike or its foundation (often visible after a flood) or of karstic activity in the substratum.

• **VEG: VEGetation** (presence of shrubs, bushes, trees or stumps).

ANOMALY CODES – LAND-SIDE PART:

• **DEP:** DEPression, pond, borrow pit (beyond the toe of the dike).

• **ISP:** Indicator of SeePage (e.g. a wet zone, seepage marks after a flood).

• **SDB:** SanD-Boils, typical circular anomalies (little sand cones or collapsed areas), visible (during or just after a flood) near to or several metres beyond the toe of the dike and indicators of the beginning of piping in a layer of the foundation.

ANOMALY CODES – RIVER-SIDE PARTS (SLOPE OR BANK):

• **FLV:** Flood LeVel / high-water mark.

• **PMC** (for the slope of the dike only): Proximity of the Main Channel (where the bottom of the true slope or toe of the dike is less than 1 metre (the convention) from the bank of the water course, irrespective of whether it has slipped).

• **RER:** Longitudinal ERosion due to the River.

• **WLV:** Water LeVel (of the river), to be recorded systematically on the day of the inspection when the watercourse touches the toe or slope of the dike.

• **DET:** DETerioration of the stones or concrete of a masonry structure, corrosion of a metal structure.

• **DIS:** DISsociation, breaking away or poor contact between 2 elements of a structure of different nature (e.g. a freeboard wall breaking away from its foundation on the dike).

• **LOO:** LOOsening, stones lost from the masonry.

• **LSC:** Loss of Structural Cohesion (in the sense of a failure affecting the structure: collapse, breaking up).

SPECIAL ANOMALY CODES:

• **PZO:** PieZOmeter device discovered or read during the inspection (recorded as an "anomaly" for the part of the dike concerned).

• **SAM:** Material SAMpled from the dike during the inspection (recorded as an "anomaly" for the part of the dike concerned, indicate in the "description" field the reason why the sample was taken and its depth in addition to the bag number in which the sample is packaged).

• **SDG:** SounDinG (in principle carried out using a manual earth auger) performed in the dike during the inspection (recorded as an "anomaly" for the part of the dike concerned; indicate in the "description" field the reason why the sounding was performed, the depth of the sounding and the reference of the geological section).

• **TPF:** Transverse ProFile surveyed during the inspection (only recorded once in the box corresponding to the part of the dike where a specific observation prompted the profiling. Indicate in the "description" field the reason why the profile was surveyed).

STRUCTURE (ELEMENT) CODES:
These codes are used either to indicate which of the dike's structural elements is affected by the anomaly concerned (in the anomaly description box), or to indicate the existence or absence of this structure on the part of the dike described (overall observations box). The structures concerned are primarily longitudinal structures, i.e. whose greatest length is parallel to the axis of the dike:

• **DIT:** DITch (or parallel drainage trench) on the land side at the toe of the dike.

• **FBF:** FreeBoard Feature, either a freeboard masonry wall or an elevation along the edge of the crest or an earth ridge, depending on the individual case).

• **PTD:** Protection of the Toe of the Dike (coarse riprap, earth berm, sheet piling or wall of wood piles), in principle on the river side.

• **PTF:** ProTective Facing of the slope of the dike (masonry stones, concrete or pre-fabricated elements).

- **ROD:** ROaD (tarred).

- **SER:** SERvice road, at the toe of the slope or on the crest.

- **SLP:** SLoPe (not faced) of the dike, on the land or river side.

- **WAL:** Support or dump WALl, on one slope of the dike.

- **WSH:** Weighted SHoulder on or banking up of the slope of the dike, on the land side or river side.

MIS CODE:
Everywhere, for MIScellaneous.

Attached:
- 1 blank anomalies record form.

VISUAL INSPECTION OF THE DIKES OF THE . . . .

Operators:

Ref. of the KM: (m): Segment_length:

Date: District: Precise location: Bank: RB / LB Page:

MM_sta: MM_end:

General positioning

Riv. side parts	Anom. Ref.	Anom. Code	Struc. Code	Nb	Long. pos	KM Calculated	Photos Nb	Photos ID	Description of the anomaly	sev. code	Overall observations: longitudinal structures/access/photos					
											Code	Y/N/U	Details about the nature of the structure			
River bank	1b										PTD					
	2b										PTF					
	3b															
	4b															
	5b										Access:		Overall view photo	Photo pos (m)	Direction	Number
	6b										Comment:				Upstream<--> Downstream	
	7b															
	8b															
River-side slope and toe	1r										PTD					
	2r										PTF					
	3r															
	4r															
	5r										Access:		Overall view photo	Photo pos (m)	Direction	Number
	6r										Comment:				Upstream<--> Downstream	
	7r															
	8r															

Crest	Anom. Ref.	Anom. Code	Struc. Code	Nb	Long. pos	KM Calculated	Photos Nb	Photos ID	Description of the anomaly	sev. code	Overall observations: longitudinal structures/access/photos					
											Code	Y/N/U	Details about the nature of the structure			
Dike crest	1c										FBF					
	2c										SER					
	3c															
	4c										Access:		Overall view photo	Photo pos (m)	Direction	Number
	5c										Comment:				Upstream<--> Downstream	
	6c															

Land-side part	Anom. Ref.	Anom. Code	Struc. Code	Nb	Long. pos	KM Calculated	Photos Nb	Photos ID	Description of the anomaly	sev. code	Overall observations: longitudinal structures/access/photos					
											Code	Y/N/U	Details about the nature of the structure			
Land-side slope and toe	1l										WSH					
	2l										DIT					
	3l															
	4l										Access:		Overall view photo	Photo pos (m)	Direction	Number
	5l										Comment:				Upstream<--> Downstream	
	6l															
	7l															
	8l															

Table of anomaly codes	All parts	VEG:Vegetation GER:Gully erosion PAS:Particular Str	MVT:Subsid/Compac/Slip/Tilting CRK:Cracking CON:Conduits/Pipes SIN:Sink hole MLE:Misc.long.str./step BUR:Burrow	Land side part: DEP: Depression/Pond ISP: Ind.seepage SDB: Sand-boils River side part: RER: River erosion FLV/WLV: Flood/water level PMC: Proximity to the main channel	Anomaly on mas/ass struct.	LOO: Loosening/lost stones DET: Deterioration stones LSC: Loss structural coh. DIS: Dissociation/diff movement	Every-where: MIS: Miscella-neous
Struc. (ele) codes	WSH:Wei.should/Banking FBF:Freeboard feature	ROD: Road SER: Service road	WAL: support wall DIT: ditch	PTD: Protect. toe dike (riprap, piling, earth berm) PTF: Protective facing SLP: Slope	Special "anomaly" codes	SAM: Sampling TPF: Transverse profile	PZO: piezometer SDG:sounding

APPENDIX 4
Presentation of an example of post-flood inspection and initial emergency repairs: visual inspection of the Agly dikes

Visual inspection of the Agly dikes
(Eastern Pyrénées "département") following flooding
on 12-13 november 1999 and repair work

1. Brief presentation of the context of the inspection

1.1 Flood protection in the lower Agly plain

The Agly is a Mediterranean coastal river with a catchment basin of just over 1,000 square km. Flooding of the Agly (always short but intense) has traditionally been dreaded by the inhabitants of the Salanque plain, lying to the north of Perpignan, (Eastern Pyrenees "Département"). The catastrophic flood of 1940 has notably remained in the memories of many people.

Before 1970, the Agly was protected by low, non-continuous flood embankments. At the beginning of the 1970s, large-scale work was undertaken to channel the river bed and build dikes on both banks for a distance of 13.2 km, running from the bridge on the RN9 (Perpignan to Narbonne road) down to the Mediterranean. Made of silty material, these dikes stand 2 to 3 metres above the plain. The crest is 8 metres wide with a road that is poorly paved on the upstream section of the diked river course and tarred on the downstream section. The total height of the dike slope on the river side is 6 metres and there is a berm halfway up. Riprap facing protects the lower part of the slope in normal sections and the whole of the slope where there are concave curves. In places, there are also concrete slabs (remains of work carried out after the flood in 1940).

The Agly dikes are built directly onto the banks of the main channel (absence of "ségonnal" or "franc-bord", discounting the small berm) and are therefore subject to considerable hydraulic constraints.

1.2 The flood of 12 and 13 november 1999

The exceptional amount of rain that fell on 12 and 13 November 1999 resulted in the Agly going into spate, the flow rate of which (peaks of 2,000 cubic metres/second recorded at Rivesaltes limnimetric station, several kilometres upstream) exceeded the capacity of the diked segment (maximum main channel discharge capacity of 1,500 cubic metres/second) downstream from Rivesaltes.

The result was that, for an hour or so, the water spilled over the top of the dikes in numerous places and on both banks, leading to miscellaneous damage, the most spectacular example being a breach roughly 50 m long in the vicinity of Saint-Laurent-de-la-Salanque's wastewater treatment plant, which was seriously damaged.

Many areas of the downstream (land side) slope were also eroded and several places showed signs of water flows in the foundations (sand-boils) or through the dike (sink holes).

1.3 Purpose of the inspection

At the beginning of December 1999, the dike operator (Pluridistrict syndicate of the lower River Agly) asked Cemagref in Aix-en-Provence to conduct a post-flood inspection of the entire dike system as well as a minimal programme of geotechnical explorations of various particular sectors. The purpose of this inspection was to obtain a rapid, comprehensive diagnosis of the structures and identify repairs to be done as a priority.

The assignment consisted of:
– A visual inspection of the left and right bank dikes downstream of the RN9 road by three operators (one of whom was working for the DDE, local Town & Country planning office) – to be carried out before the end of December 1999.
– The recording of observations on report forms and maps, as well as the production of a photographic dossier.
– Geotechnical explorations of certain particular anomalies areas (by sounding and soil samples obtained by mechanical shovel) followed by soil mechanics tests in the laboratory.
– The drafting of a synoptic report specifying the repair and upgrading techniques to be employed and any further investigations that may be required.

2. Post-flood visual inspection of dikes

2.1 Method used

Post-flood visual inspections were conducted on the left bank on 14, 16 and 17 December 1999 and on the right bank on 15, 29 and 30 December 1999. They were carried out by three operators, two of whom had a notebook and camera. A further inspection was conducted on 9 and 10 March 2000.

POSITION FIXING

Positions were fixed using a Topofil thread-measuring device, starting from the downstream facing of the RN9 bridge abutments (MM 0). The measurements obtained with the Topofil were regularly re-positioned against fixed points (crossroads, field boundaries, etc.), which meant that any necessary corrections could be made once back in the office using information on the available 1:5,000 map.

Two position reference systems were produced in this way (for the left and right banks) that were used for the entire study.

Checks made during subsequent visits estimated the accuracy tolerance for the measurements with respect to the bi-kilometric markers indicated on the 1:5,000 maps to be between 5 and 15 metres, taking into account all sources of error and depending on the distance between the identified point and the crest. Nowadays, we can obtain the same precision more simply with a pocket GPS receiver, if such a device can be used (no dense forestry cover).

DISTRIBUTION OF WORK

The person working on the dike crest was responsible for mapping the position using the Topofil and for reporting anomalies and deterioration on the top of the structure, the road and its verges.

A second operator was responsible for visually inspecting the river side slope and bank.

The third member of the team worked on the description of the land-side slope, noting any information about the crest provided by the person working on top of the dike.

Roughly 4.5 to 5 km of dike were covered by the three-person team each day.

Information was transcribed back in the office (descriptions, photo captions, cartography) at a rate of 3 to 4 km a day by two of the team (one dealing with data concerning the river side and the other with data for the land side and crest).

RECORDING OF ANOMALIES

Inspection findings were written out in a linear manner in the form of tables (see abstracts in section 2.2).

A given table corresponds to a date-segment (portion of dike covered on the day indicated), to a three-person team and to one of the two dike slopes visited.

The name of the team member working on the slope described appears in bold print at the top. The date of the visit and the metre markers (MM start to MM finish) are also given for each segment.

Example of report headings *(for the description of a land-side slope)*

14 December 1999:   River-side operator: Cyril Folton (Cemagref)
MM: 0 to 3850   **Land-side operator: Sébastien Villa (DDE 66)**
      Crest operator: Sébastien Merckle (Cemagref)

Metre markers	Description	Photos	Photo code*

– The metre marker (MM) is taken to be the point (or segment) described and obtained with the Topofil after correction in the office.
– The description covers the nature of the slope or of the anomaly(ies) noted.
– The line appears in bold print when the description or anomaly involves a homogeneous segment.
– The description of particular features, whether one-off or not, within a homogeneous segment appears in normal print in the table.

PHOTOGRAPHS

Three types of photographs (totalling 400 in all) were taken:

a) Digital photos, with a reference N° such as "LBLand-MM7058.d", with the following meanings:
– 'LB' for left bank and 'RB' for right bank.
– 'Land' for a photo concerning the land side.
– 'Crest' for a photo concerning the crest.
– 'LB' or 'RB' alone for bank-side slopes.
– MM7058 indicates the metre marker from which the photo was taken.
– 'd' for digital. '(2) d' or '(3) d' means that there are 2 or 3 shots of the same view.

b) Slides, with a reference N° such as "MM 8598 - 3 S":
– '3' for the N° on the slide.
– 'MM 8598' for the metre marker where the shot was taken.
– 'S' for slide.

c) Printed photos, with the same type of reference N° as that used for slides, followed by 'P' for printed. Photos taken on subsequent visits on 19 & 20 January and 9 & 10 March are indicated by the date in brackets.

Any mention of "upstream" or "downstream" after a photograph's metric reference (e.g. MM 8598 – upstream) indicates that the photo was taken from that MM looking upstream or downstream of the River Agly.

CARTOGRAPHY

The main observations derived from descriptions were transcribed onto A3-format boards (copy of 1:5,000 maps from July 1993). The key to maps is the following:

Slope protection	Continuous	Discontinuous or uncertain
Riprap at toe of bank (below the berm)		
Riprap on the whole bank (including dike slope)		
Recent riprap repairs (sometimes masonry blocks)		
Concrete wall or protection		
Recent groynes (riprap)		

Trees	Isolated	Fallen or uprooted
Trees		
Clumped trees		

Deterioration	Isolated	Widespread
Scouring or erosion of the base (foundation anchor block damaged if one exists)		
Burrows or warrens		
Soil creep/landslide		
Localised erosion		
Sand-boils, artesian upwelling near land-side toe of dike		
Presumed site of overtopping		

Other data appearing on maps:
– Conduit crossings.
– Bi-kilometre markers for the position reference system.
– Position of the main channel at the time of inspection.
– Main channel or tributary channel in contact with the toe of a bank or anchor block.

2.2 Extracts from data collection tables (given as an example)

Post-flood visual inspection of the left-bank dike of the River Agly
A. River-side slope

The sector described is the left bank of the River Agly downstream of the bridge on the RN9 road to the sea. The datum point for metre markers was the left-bank downstream pier of the bridge.

14 December 1999:   **River-side operator: Cyril Folton (Cemagref)**
MM: 0 to 3850   Land-side operator: Patrice Mériaux Cemagref)
    Crest operator: Sébastien Merckle (Cemagref)

• **Photo codes:** D for digital photos; S for slides and P for prints.

Metre markers	Description	Photos	Photo code*
0	Bridge on the RN9 road		
0 to 90	The bank is divided into 3 areas: • The top of the bank is steeply sloped, approximately 2.5 m high & covered with giant reeds. • A 3 m wide horizontal berm is protected by large pieces of riprap. The lower slope is also protected by riprap covered in silt deposits and overgrown with giant reeds.		

20	Disorganised riprap on the berm	LB-MM20 upstream	D
70	Widening of the "ségonnal" at the toe of the dike (5 to 10 m)		
90	End of riprap at top of bank and appearance of burrows. Extensive silt deposits at base of bank	LB-MM90 downstream	D
120	Trees on berm		
150	Riprap visible on lower bank		
190 to 380	Warrens, disorganised riprap on lower bank; reeds on upper bank and berm	LB-MM190 Burrows	D
229	Uprooted tree at toe		
260	Presence of moles		
275	Typical profile: berm riprap and giant reeds	LB-MM275 downstream	D
380 to 551	Trees and shrubs on the berm. "Ségonnal" width approx. 20m	LB-MM380 downstream LB-MM495 bank	D
390	Fox's den or badger's set at 1m below crest	LB-MM390 burrow	D
475	Badger's set	LB-MM475 burrow	D
486	Burrows at 1.5 m below the crest		
494	Burrows		
502	Collapse of old burrows		
508	High-water marks at 0.50 m below the crest		
512	Burrows		
555 to 620	Moles. The upper bank is covered in reeds and the lower bank is protected by smaller riprap (0.40 m)	LB-MM550 Downstream bank	D
605	Burrows, 40 cm in diameter		
620 to 837	Riprap on whole bank, trees on berm and reeds	LB-MM620 Riprap	D
670	Disorganised riprap on berm		
700	Riprap larger than 1,000 mm and over		
742	Disorganised riprap, high-water marks in trees on the berm and obstructions/log jams	LB-MM742 Riprap and high-water marks	D
825	Rainwater discharge. "Ségonnal" widening to almost 30 metres	LB-MM825 Rainwater discharge	D
Etc.	Etc.		

Post-flood visual inspection of the left-bank dike of the River Agly
B. Land-side slope (and crest)

The sector described is the left bank of the River Agly downstream of the bridge on the RN 9 road to the sea. The datum point for meter markers was the left-bank downstream pier of the bridge.

14 December 1999:   River-side operator: Cyril Folton (Cemagref)
MM: 0 to 3850   **Land-side operator: Patrice Mériaux (Cemagref)**
Crest operator: Sébastien Merckle (Cemagref)

Photo codes: D for digital photos; S for slides and P for prints.

Metre markers	Description	Photos	Photo code*
0	Bridge on the RN9 road		
0-275	Dike land-side slope, the height of which increases gradually from 0.8 m to 1.4 m, from upstream to downstream. Mostly gentle slope (under 50%). In places, clumps or lines of reeds, sometimes scrub and undergrowth. Service track at toe.	MM0-downstream MM275-upstream General views	2 P 3 P
0-35	Overgrown slope (including a few young bushes). Scour hole at downstream outlet of ARMCO pipe under RN9 road (in principle due to water coming from upstream of the road). Track at bottom of dike almost joins crest dike. Minor slope erosion (stony ground) in the vicinity of an EDF transformer. Rain gauge at MM25 on the land-side edge of the crest.	MM35-upstream Ditch at ARMCO pipe outlet	0-1 P
60-80	Line of shrubs on lower slope.		
90	Connecting path between tracks at toe of slope and on crest.		
90-140	Giant reeds on the slope. Slope height approx. 0.80 m.		
140-190	Slope height rising to 1.4 m. Line of reeds on crest edge where the slope becomes slightly steeper.		
190-210	The slope gets gentler.		
210-275	Few or no reeds.		
275-475	Slope with reeds, steep in places with height around 1.5 m. Service track at toe.		
298	40 cm diameter burrow half-way up slope (sandy soil, rabbit droppings).		
340	Gradient of slope – 53% (clinometer). Still reeds.		
350	Steep gradient – 70-80%. Height – 1.5 m.		

400	Old burrow, 20 cm in diameter, on lower third of slope.		
440	Beginnings of 20 cm diameter burrow (silty soil).		
450	Beginnings of 10 cm diameter burrow, on lower third of slope.		
400-450	High-water marks visible among vines on land side.		
475	Small ashlar construction (presence of an air vent and electrical supply) on east foot of electricity pylon. The diverted end of the track cuts slightly into the toe of the dike.		
475-810	Slope height 0.80-2.0 m with more or less dense reed cover. Absence of service track at toe of dike.	MM600-upstream General view	4 P
485	Gradient – 50%. Height – 1.8 m.		
660-680	High-water marks at toe of slope – 0.3 to 0.4 m above natural ground. Rabbit droppings on ground.		
670-700	Slope height –1.7 m.		
752	Slope height – 1.5 m.		
750-780	Beginnings of rabbit warrens at toe of slope.		
810-1030	Lower slope – 1.4 m to 1.2 m. Reed cover varies. Still no track at toe.		
810	Transverse track reaches the crest. Cast-iron inspection hole cover on track beyond toe of dike. Watering hole a little further into the field (well? + pump).		
Etc.	Etc.		

3. Diagnosis-inspection report and suggestions for repairs to be done (summary)

Post-flood visual inspection of the Agly dikes, together with a number of exploratory boreholes made with a mechanical shovel, made it possible to locate, list and describe the main damage to structures (dikes and bank protection) following events on 12 and 13 November 1999. It was used to work out the emergency repairs and upgrading work needed (most of which was carried out in the spring of 2000).

3.1 Sectors of dike overtopping

On the left bank, areas where overtopping had occurred (including the breach by the waste water treatment plant in St-Laurent-de-la-Salanque) were repaired in the days following the flood peak.

On the right bank, visual inspection made it possible to catalogue sectors where overtopping had caused serious erosion (rill erosion, niches, soil creep and mudslides) on the dike's land-side slope. This largely concerned three areas in the downstream section of the diked river course (beyond metre marker 9500) for a total distance of around 550 m.

It was an urgent matter for these seriously eroded areas to be repaired since they represented potentially vulnerable points in the dike: very susceptible to erosion especially if overtopped again (concentration of water flows due to excavations and unevenness), instability of subvertical gradients that had developed, and reduction in seepage path length (risk of internal erosion due to the hydraulic head). The subject of resolving, for the long term, the problem of overtopping is not dealt with here (see section 3.7).

Suggested solutions and their implementation:
Construction of a compacted earthfill embankment with foundations drained by a granular blanket, with a view to arriving at a land-side slope with good mechanical properties and providing adequate drainage of this part of the dike.

3.2 Seepage and pervious dike body

The post-flood visual inspection revealed evidence of seepage having surfaced on the land-side slope at a precise metre marker on the dike's right bank (MM 1720). Further exploration with a mechanical shovel provided the explanation for the leak – a bank of large riprap inside the dike body – and revealed the upstream-downstream distance covered by the anomaly (about fifty metres).

The principal risk created by this dike's heterogeneous and pervious nature was that of internal erosion (piping), especially in transition areas between riprap and silts, the number and geometry of which was not known. In fact, this sector of dike appeared to be abnormal for a distance of more than 400 m because it encompassed an old concrete wall within its river-side slope (slope protection situated on the outside of a bend in the river and probably built after the 1940 flood), behind which other indicators of piping were noted, such as sink holes.

Suggested solutions and their implementation:
To make the dike sufficiently impervious and in view of the heterogeneous nature of the embankment, the suggestion was made to replace the old concrete wall by a supportive, impervious and protective structure made of large, carefully cemented riprap.

3.3 Erosion of the foundation anchor block at the toe of banks

In 1970, the design model profile used for dike construction along the River Agly anticipated the systematic inclusion (along the whole dike) of a foundation anchor block made of large riprap (300 kg to 2.500 kg), 3.5 m wide and 1.2 m deep. Generally speaking, the visual inspection of December 1999 showed that, in the places where this foundation should have been visible (e.g. at the toe of a scoured bank), it was effectively there and its width was frequently greater than that planned (up to 5 m).

In some segments, which were identified on maps, this anchor block had been attacked during the November 1999 flood and parts of it (width) swept away. The phenomenon affected a total of 1.4 km on the right bank and 1 km on the left bank. In places, the scoured foundation was in direct contact with the current active bed of the Agly, which aggravated the situation since there was a risk of scouring making deeper inroads during periods of minimum flow and regular flooding. However, nowhere did the bank protection itself seem to have been affected to any great extent, which pointed to the relatively satisfactory hydraulic behaviour of these protective arrangements, even when main channel flow rate was at maximum capacity.

The fuse plug – which, in a way, was the function of the foundation anchor block – thus worked well in those places where it proved necessary and it would be appropriate to rebuild it to do the same job in the future.

Suggested solutions and their implementation:
Reconstruction of the foundation anchor block using large, dry (not cemented) riprap, at least to the same dimensions, and backfilling of scour holes.

3.4 Lifting due to groundwater pressure (sand-boils)

During post-flood visual inspection, it was noted that two specific sectors of the dike (one on each river bank) had a band of twenty metres or so along the toe of the land-side slope where unusual shapes had formed – either craters or collapsed areas 0.5 to 2 m in width or very regular circular mounds (cones) up to 1 m in diameter. In international literature, these particular phenomena are called "sand-boils". Using a mechanical shovel, it transpired that these shapes had appeared only in areas where a layer of silts covered the alluvial gravel of the dikes' foundations.

The explanation was that, because of the presence of this relatively impervious layer, the Agly's accompanying water table was under pressure when the river was in spate and increased the head behind its dikes. A pressure gradient was thus created between the base of the silt layer and its summit. If the silt layer is thick enough, the "cover is kept tightly in place" and nothing happens. If not (thickness threshold of 2 metres or so?) and because of the effect of the pressure that tries to escape, air is dispersed (explaining the bubbles reported by witnesses), as is water and probably soil particles, through the silt layer leading to the formation of these convex (cones) or concave (collapsed areas) shapes.

The main risk that these phenomena represented for the dike was that they occurred on or in the immediate vicinity of the toe of the land-side slope, the stability of which could thus be compromised.

Suggested solutions and their implementation:
Creation of a decompression device using a drainage ditch with a maximum depth of 1.5 m, running along the toe of the dike at a distance of at least 2 metres (to avoid

destabilising the toe when the temporary excavation is made) and designed to draw off uplift pressure where it was likely to have a negative impact on the dike; that is, near the toe of the land-side slope.

3.5 Longitudinal cracks in the dike crest

As in the wake of flooding in October 1992, longitudinal cracking was noticed (this time in two sections of the right-bank dike crest), the position and shape of which (arc of a circle) could point to the beginnings of soil creep on the river-side slope. However, in contrast with damage in 1992, the 1999 post-flood visual inspection did not reveal any tangible sign of slope deformation (e.g. hummocky areas on the slope or a swelling on the berm) in the vicinity of these two areas of crest cracking. It should be said that, in both cases, the surface of the river-side slope was seriously affected by numerous burrows, which hindered observation and were also doubtless partly responsible for the damage (having encouraged the infiltration of water inside the dike body during the flood).

Although they did not create breaches, these ground movements compromised the geometry and integrity of the dike.

Suggested solutions and their implementation:
Since stabilisation work on the three areas of soil creep resulting from the 1992 flood proved to be effective (no new damage to the consolidated sections reported during the inspection that followed flooding in November 1999), it was logical to suggest that similar methods be used to consolidate the two new areas of cracking – purging of the undermined and potentially unstable materials on the river-side slope and berm (if not riprap) and replacement with a stretch of large, dry (not cemented) riprap material.

3.6 Miscellaneous damage

Other more occasional or theoretically less worrying damage was diagnosed during the post-flood inspection:
– Areas of suspected seepage: witness accounts to be obtained and surveillance to be carried out during flooding.
– Evidence of burrowing animals, mostly rabbit warrens, the entrance to which was sometimes made bigger by dogs, although significantly affected areas were relatively infrequent. Proposed solution: backfilling and riprap facing (when associated with slope consolidation) or installation of preventive metallic netting after re-profiling.
– Vegetation was poorly maintained on the dike's slopes. It therefore impeded visual surveillance (giant reeds), made for a quiet life for burrowing animals and was potentially directly responsible for damage (uprooted trees). On the other hand, bushes and trees would break up the monotony of the artificial environment of the dike-protected Agly river. Besides which, the dike crest was wide (8 m or so), which reduced the threat from piping caused by rotting tree roots. The suggestion was therefore to opt for annually mown herbaceous cover on the dike slopes themselves and to draft and put into practice a concerted plan for managing bank and canalised river bed vegetation.

– Sink holes: Indisputable evidence of sink holes on parts of the dike crest appeared several days after the flood peak and some continued to develop in the following weeks. They generally appeared as the result of heterogeneity in the dike body. Recommendations were made for close surveillance during and after flooding and/or for work to be done to improve dike impermeability.

Finally, recommendations were made to the syndicate (director of works) to equip themselves with effective mapping resources for dike maintenance purposes, which would make it possible to detail, on a daily basis, the nature and location of repair and maintenance work carried out. Whilst waiting for such a resource, which was tragically lacking, a request was made for site-by-site technical reports and drawings that actually corresponded to the reality of structures (especially for work sites started in the wake of the 1999 flood).

3.7 Towards dealing with the risk of overtopping

During the flood of November 1999, overtopping occurred because flood-water discharge exceeded the capacity of the diked bed of the Agly.

Since river-bed reshaping and dike construction in the mid-1970s, the developed section of the Agly had been subject to a maximum main channel discharge (flooding in October 1992) and an overtopping discharge (flooding in November 1999). This corroborated the fact that the dikes were designed to withstand twenty to thirty-year floods.

Whatever the exact figure, it was obvious that overtopping would occur several times each century and that, in the absence of spillways[3], would again result in breaches[4] that would be impossible to predict and lead to:
– "At best", damage to the dikes themselves: breaches and/or erosion of the landside slope, which would need to be repaired as soon as possible after overspill flooding.
– "At worst", damage to buildings or developments located in the vicinity of the downstream toe of the dike in a breached area (the breach at St Laurent illustrated this scenario, having caused severe damage to the wastewater treatment plant), without mentioning the potential threat to human life if any of the buildings under threat were lived in.

The dike system of the lower plain of the River Agly therefore needed to be developed to give it a safety margin to enable it to withstand, with no damage, floods with a typical recurrence interval of well over thirty years. The aim would not be to increase the capacity of the diked channel, but to prevent – or, more exactly, to delay – the formation of sudden breaches in floods that exceed the capacity of the diked system

---

3. Lower segments of dike that have been built (stone or concrete facing with a runoff invert) to allow water to spill over without causing damage to earthfill sections or the bottom of slopes.
4. The hypothesis that provides the grounds for this statement, which is only too frequently confirmed by the facts, is that earthfill hydraulic structures (dikes, but also earthfill dams) do not withstand overtopping.

(30-year flood peaks). This called for the building of safety spillways, but only after a full topographical, hydraulic and flood probability study had been conducted and after negotiations aimed at identifying spillover sites where the socio-economic impact downstream would be the least in view of current land use.

Such a study was conducted in 2001-2002.

APPENDIX 5
Annotated digest of the main French regulations concerning flood protection dikes

(1) Interministerial circular dated 24 January 1994 on flood prevention and the management of areas liable to flooding.

In this text, the French government takes a deliberately harder line concerning areas liable to flooding, based on three principles: the prohibition of all new building in areas liable to flooding, strict controls on further urbanisation in flood storage areas, limitation of new dike systems and aggradation.

Mapping of areas liable to flooding (atlas of flooded areas, PER, PSS, R111-3 map, etc.) constitutes the priority resource for implementing this policy.

(2) Interministerial circular dated 17 August 1994 on procedures for flood protection works management.

Following the Camargue floods during the winter of 1993-1994, this is the first text to require Prefects to draw up an inventory of dike operators and, if possible, of the structures themselves.

(3) Circular DE/SDGE/BPIDPF-MPN/N°629 dated 28 May 1999 issued by the Ministry of the Environment on the subject of cataloguing dikes that protect residential areas against inland waterway and coastal flooding.

Launch of a national inventory of dikes, their operators and protected areas, in association with France's "DIGUES" software.

(4) Decree N°2002-202 dated 13 February 2002 in amendment of decree N°93-743 dated 29 March 1993 on the subject of the nomenclature of operations requiring authorisation or declaration in application of article 10 of Act N°92-3 dated 3 January 1992 on Water (Official Journal of 16 February 2002 and MATE official bulletin of 21 May 2002, p. 49-50).

This decree incorporates the following section into the water policy nomenclature: "2.5.4. Installations, structures, dikes or embankments with a maximum height greater than 0.50 m above the natural ground in a watercourse's floodplain". These civil engineering installations now require preliminary administrative authorisation or declaration depending on surface area and/or the width they occupy in the floodplain.

(5) Circular MATE/SDPGE/BPIDPF/CCG N°234 dated 30 April 2002 on the subject of government policy concerning predictable natural risks and the management of areas situated behind flood protection dikes and maritime flooding.

Summary of the principles of government policy on the risk of maritime flooding and inundation, and formulation of a position on urbanisation in dike-protected areas, notably within the framework of drawing up action plans for the prevention of flood risks ("PPR-I" in France).

(6) Interministerial circular DE/SDGE/BPIDPF-CCG / N°8 dated 6 August 2003 issued by both the French Ministry of the Interior and the Ministry of the Environment on the subject of organising the Water Policy Control of flood protection dikes, the failure of which may threaten "public safety".

Following the example of dams in 1970, this circular introduces a policy control system for dikes the failure of which may threaten "public safety", with a definition of the obligations of owners on the one hand and of controlling bodies (the Water Police in this case) on the other. Structures concerned are dikes that protect against watercourse spillover, including torrents, as well as dikes built in connection with the 'dynamic slowdown' of developments.

(7) Interministerial circular-letter dated 21 January 2004 issued by the Ministry for Town and Country Planning and the Ministry of the Environment on the subject of managing urban development and of the adaptation of buildings situated on land liable to flooding, for the attention of the Prefects of the nine following 'départements' situated in the French Mediterranean region: Ardèche, Aude, Bouches-du-Rhône, Drôme, Gard, Hérault, Lozère, Pyrénées Orientales and Vaucluse.

With the feedback obtained from flooding in the last ten years, implementation of an action plan relating to the management of urban development on land liable to flooding, based on four themes:
– Control of urban development.
– Adaptation of buildings.
– Management of protective structures.
– Organisation of actions and resources.

In relation to the "management of protective structures", this text recalls the necessity not to increase vulnerability behind dikes and to take into account the hypothesis of the failure of protective structures during floods.

Imprimé pour vous par Books on Demand (Allemagne)